TECHNOUTOPIA

How optimism ruined the internet

Second edition
Copyright © 2018 Alex Warren
All rights reserved.
First published: December 2016
ISBN-10: 1539567540
ISBN-13: 9781539567547

Contents

Preface i

Introduction 1
An argument for nuance   1
The internet wants to be free markets   6
The age of technoutopia   9
Returning to realism   13

The internet encourages democracy, equality, and transparency 16
Paying for a megaphone   22
We are the 0.01%   28
The failure of journalism 2.0   33
The success of spin 2.0   44
The internet will find a way   55
A change in our thinking   73

The internet removes bureaucrats, intermediaries and gatekeepers 78
Gatekeepers: An end to the elite   82
An attack on inefficiency   86
Amazon – A new generation of gatekeeper   95
Search engines: Organising information, on your behalf   105

Calling Google to account   110
Scraping the content barrel   115
Lifting the Google Lens   120
Opening the gates   126

The internet improves our ability to communicate and socialise   129
The online world is flat. You are flat   133
#Awkward   139
Brand Me   144
Downloading selfishness   149
An unconformable place to live   153

A realist conclusion   156

References   162

For Jam

Preface

It's now been nearly three years since I finished writing Technoutopia, and somehow the world feels like an entirely different place.

When preparing the first edition in 2015, technological utopianism was at its peak. There was no Trump, no Brexit, no Cambridge Analytica. Uber democratised transport on the streets of London, while self-driving cars moved freely throughout Silicon Valley. Elon Musk was taking us all to Mars and the miracle of social media was still poised to bring humanity closer together – rather than tearing it apart.

On writing the first edition of Technoutopia I could not have chosen a timelier, yet more controversial subject. Technologists, futurists and industry analysts were quick to disavow the central tenets of the book, claiming that an attack on what they deemed to be 'inevitable progress' was simply unhelpful to the overall debate. In one surprisingly honest review, I was informed by a futurist that "I cannot be seen to promote something that attacks the industry which pays my bills". This sentiment goes a long way in describing the unhealthy and self-reinforcing attitudes of the technology industry at the time. In a world built on positive visions and optimistic TED Talks, to be critical was to risk losing your appeal and possibly even your livelihood.

Now, three years later, we are living in a whole new – and more complicated – world. Uber has been banned from major cities around the globe, the self-driving cars of Silicon Valley have claimed their first human life, open information has given way to an endless stream of fake news, while online advertising has not only come to jeopardise the integrity of the media, but is now undermining democracy itself.

The technoutopianism that came to inform the previous five years is dead, but the thought processes that allowed it to arise in the first place are still very much alive within the technology community.

Following the string of scandals that have engulfed many of the world's largest tech companies in 2018, Silicon Valley has been keen to prove that it is taking issues of privacy, information control and democratised technology seriously. To achieve this, these businesses have turned to the only solution they understand – technology.

Rather than accepting the reality that vastly complex social issues require time, thought and cross-society collaboration to address, the technology giants have been keen to prove that these issues can be solved almost overnight – and without the need for time-consuming academic research or costly government intervention. With the exception of clearer privacy settings, which are frankly long overdue, almost all of the technological solutions provided have either proved to be a disaster, or have opened up a whole host of new and unexplored ethical concerns.

Facebook's suggestion that it could combat fake news through artificial intelligence was roundly mocked, with even the most basic trials proving that AI (or in this case, machine learning) is simply incapable of the nuanced decision making needed to distinguish reality from fiction. While I'm sure artificial intelligence will reach a point where it can differentiate between the two, the feasibility of such technology is not the issue.

Do we really want a society where a piece of code decides which information sources we do and do not see? Even more importantly, do we really want to live in a society where that piece of code is owned, built and operated by a private corporation?

This is the thought pattern that Silicon Valley simply doesn't seem capable of overcoming, that the solution to this problem may not come from them. Even more importantly, such a change in mindset would mean admitting that not all of society's problems are solved by greater investment in internet technologies.

Instead of realising this however, we continue to see blunt, technological solutions applied to problems that originally arose as a result of blunt and ill-planned technological adoption.

And it's not just Facebook that has struggled to let go of its tech-centric mindset. In the fight against online abuse, both Twitter and YouTube have grown ever more absolutist in their crack down on hate speech. While I wouldn't advocate many of the views being censored by these networks, it's important to ask the question of whether we want a corporation deciding which political views are and aren't acceptable in the sphere of public debate – especially when these same platforms fail to provide an exact definition of what is and isn't considered hate speech. Yet again, we are faced with corporations trying to solve a problem without relinquishing any control over the issue. If anything, their solutions result in an *increase* in control.

While it's all too easy to attack these technologies in the current climate, the fact is that governments and democratic institutions have done an equally bad job of finding solutions to these fast-moving societal issues. Where Silicon Valley seems keen to fix the internet overnight, governments across both the UK and the US are still struggling to even understand what the issues are, let alone how the proposed technological fixes could apply.

Nowhere was this clearer than in the testimony of Mark Zuckerberg before the US Congress. Following the Cambridge Analytica scandal in which illegally sourced Facebook data was allegedly used to influence the 2016 presidential election, Zuckerberg was called before Congress for a grilling about Facebook's practices and potentially unethical use of customer data. Even as someone who strongly believes that government still has a role to play in our future society, it's hard not to accept that this process was an embarrassment to all those involved. Not only were the senators poorly placed to question Mark Zuckerberg about Cambridge Analytica, many of them seemed to lack even the most basic understanding of how the modern internet works.

Even twenty-year-old concepts, such as advertising-funded business models, seemed to bewilder the panel, rendering much of the interrogation more of a Q&A session for the technologically illiterate. As a result, any regulatory propositions to fall out of this process are likely to be equally ill-informed and ultimately ineffective.

While this may sound like a pessimistic outlook, the reality is that positive steps are being taken. Even the fact that governments and technologists are discussing these issues and thinking about potential solutions is a significant leap forward for those who, only three years ago, were blinded by technoutopianism.

Admitting we have a problem has been the first step. The next will be realising that no one solution is going to emerge overnight – and it's certainly not going to come in the form of a technological miracle cure. Addressing the issues of consumer privacy, of oligopolistic information control and of technology dependency will not be a quick process, nor will changing the solutionist mindsets of those embedded within the technology industry.

What we need more than ever is a nuanced approach. We cannot go on offering technology unquestioning praise, but we also cannot risk throwing the baby out with the bath water. The pendulum must settle somewhere in the middle, and to achieve that we need to adopt a technorealist mindset.

While many of the technological examples referenced throughout this book may seem antiquated three years down the line, the mantra of technorealism demanded throughout feels more important than ever before.

Despite being developed in the mid-1990s, the manifesto of technorealism summarised in the opening chapter of this book still stands today. The idea that 'information freedom' is not an argument for free market economics (but is instead a mantra for human-centric design) is vital in finding an authentic fix to the digital world. At the same time, we must also come to understand that calling for democratic responsibility is not always the same as calling for increased regulation or greater government intervention. To my mind, the future of a healthy internet will never be achieved by blocking and banning, but rather by pausing and thinking. It will be achieved by collaborating and thinking carefully about what the right questions are to ask, and who would be the best people to answer them.

We are already seeing a radical demand for increased diversity across the tech sector, with both technology firms and non-profit organisations calling for a wider array of representation for ethnic and minority voices. While it's important that these voices are incorporated organically, this takes us back to the central lesson of technorealism – that the best results rarely come from a quick-fix solution.

To address the social biases inherent in many of today's technologies, we must learn to look beyond the technologies themselves. We must question why the internet has been structured in the way it has, and which organisations are helping to ensure that it remains structured in that way. Next, we must look beyond the companies themselves, examining how their organisations are structured, what their corporate cultures are derived from and who has been instrumental in defining these cultures.

Even beyond this, we must look to the talent pool and ask ourselves how much choice was available to these organisations when building their teams. Do they even have a chance to reflect multiple viewpoints or is the available talent so limited to one particular geography, culture or worldview that the resulting structures feel almost inevitable?

This is the level of scrutiny that technorealism demands of us, looking beyond outright criticism of technology, and incorporating a rigorous scrutiny of the entire ecosystem within which that technology has been built. It is only by examining technology at this level – at its social infrastructure – that we can begin to rebuild from the ground up.

For our generation, this may all be too little too late. But, if we act now, we can instil in the next generation an entirely different attitude towards technology, an attitude that will one day define the next era of human-centric disruptive tech.

For me, this journey has to start in schools. The lack of focus on media literacy and how the tools we use every day impact our beliefs represents one of the greatest failings of the modern education system. In a world where children as young as four have access to mobile phones, tablet computers and the unrestrained internet, the critical study of these technologies must be a priority. In studying these tools, we not only hope to encourage a more nuanced view of technology, but also to inspire greater interest in helping to solve its consequences in future. In the process, we will inevitably expand the pool of people looking to pursue a career in tech – something which, to my mind, can only be a good thing.

As well as addressing this issue in schools, parents should also work hard to instil a sense of technological autonomy in their children – to help them understand that technology is not something that just happens, but something that can be defined and shaped by them. In the age of smart homes and the 'internet of things', this realisation is more important than ever.

Where once we worried about children using social networking sites, today's young people will be raised in an environment of near constant surveillance and personal data collection. For every smart device that gets added to our homes we are essentially installing another point of data collection – all without the express consent of those living under our roofs.

If we are to thrust such invasive technologies upon the next generation – again, something which is entirely up to us – we should at least provide them with the critical faculties needed to understand their circumstances and to question whether the current system is the best we can do.

While this may all sound like a faraway future, it is our responsibility to shape that future, and ensure that our decisions reflect not only what is in our best interest, but is in the best interest of the next generation.

This is the key lesson to take away from the last few years of technological disruption – that in order to fix the internet, we must be willing to take on an element of personal responsibility. As we have already seen, left to their own devices, tech companies will not always do what's best. Provided with free reign, Silicon Valley's mantra of open disruption ended with job losses, political manipulation, sweatshop production lines and some of the fastest-growing wealth divides in global history. At the opposite end of the scale, the ham-fisted efforts of ill-informed governments have led to innovation-strangling regulations and a society that limps forward when it could easily sprint.

What we need is a combined solution, one in which governments and corporations work together to empower consumers – designing the best technology and knowing how to implement it with minimal disruption to society. Laws and regulations will remain a fundamental part of this process, but instead of being seen to restrict innovation, they should be viewed as a way to protect consumers in the long term.

Achieving such a balance will be a slow process, but perhaps that is precisely what we need right now.

Silicon Valley's famous culture of 'move fast and break things' is broken; now all of us in the technology community must focus our energies on moving slowly and fixing things.

Introduction

"It is a mistake to suppose that any technological innovation has a one-sided effect. Every technology is both a burden and a blessing; not either-or, but this-and-that."

Neil Postman

An argument for nuance

The year was 1996. The desktop computer was a functional luxury, capable of sending emails and very little else. Nerds bragged about their 16MBs of RAM and their colour monitors (if they could afford them). The palm pilot and the CD-RW hit the UK shelves, with one starry-eyed technologist correctly predicting that "DVD is going to be very big. Limited only by the imagination."[1]

In the business community, a new online bookstore called Amazon was beginning to garner attention. In Silicon Valley, there was a buzz building around the newly launched Hotmail platform, which promised freedom from centralised ISP email clients. Deep in the Computer Sciences department of Stanford University, two PHD students theorised over the concept of "PageRank" as a new way to organise the world's information. Little did they know that their invention would one day change the way that an entire species would learn, communicate and evolve.

In the midst of all this progress, a small group of academic dissidents turned against the tide and began to question the direction that these seemingly miraculous technologies were taking them in. At the heart of this movement was a new concern for how technology was likely to erode existing cultures, and how an abundance of information could one day lead to a hyper-informed, yet under-educated society. In this society, the new generation would develop data without knowledge, would experience communication without companionship and would ultimately come to see the advancement of technology as an end rather than a means to progress.

Inspired by the writings of Marshal McLuhan and Neil Postman, these concerned academics, technologists and journalists came together to flesh out a plan for how existing cultures could fight back – or at least stand their ground – against the new wave of technological "progress". Faced with unquestioning optimism towards technology, this small group of dissidents needed a term that defined their goals. To call them Luddites would be to fundamentally misunderstand their cause. They did not call for a reversal of progress or even a greater degree of government control, instead, they simply requested a degree of realism; an opportunity to collectively buffer as a society, giving time to consider whether the negative consequences of technology could ever outweigh the demand of the market. It was in this quest for nuance that they found their new epithet: Technorealism.

Described by one of the movement's founders, Technorealism was "neither technoutopian nor neo-Luddite", but rather an attempt to "appreciate the benefits of technology while recognising and responding to its drawbacks."[2] It was a call for nuance, for an understanding that just because something is new or progressive, does not mean that it is above scrutiny or critical analysis. As the American lecturer David Shenk explains in his manifesto to the cause, *Data Smog*, Technorealism is:

> "An attempt at a more balanced response to the technological revolution, appreciating the benefits of technology while keeping a careful eye out for the drawbacks. Technorealism rejects the notion that

cyberspace or any technology is a truly autonomous world unto itself, exempt from all laws and social conventions." 2

Armed with this basic principle, the technorealists set out to define the key tenets that they, as professional technologists, would aim to uphold. Outlined in a fantastically nineties website (which can still be viewed online today at *www.technorealism.org*[3]) these progressive rules set the tone for a new era of internet discussion, combining technological optimism, political nuance and a degree of cultural caution that was previously unheard of in internet debate. They were:

1. Technologies are not neutral.
A great misconception of our time is the idea that technologies are completely free of bias – that because they are inanimate artefacts they do not promote certain kinds of behaviours over others. In truth, technologies come loaded with both intended and unintended social, political, and economic leanings. Every tool provides its users with a particular manner of seeing the world and specific ways of interacting with others. It is important for each of us to consider the biases of various technologies and to seek out those that reflect our values and aspirations.

2. The Internet is revolutionary, but not Utopian.
The Net is an extraordinary communications tool that provides a range of new opportunities for people, communities, businesses, and governments. Yet as cyberspace becomes more populated, it increasingly resembles society at large, in all its complexity. For every empowering or enlightening aspect of the wired life,

there will also be dimensions that are malicious, perverse, or rather ordinary.

3. Government has an important role to play on the electronic frontier.

Contrary to some claims, cyberspace is not formally a place or jurisdiction separate from Earth. While governments should respect the rules and customs that have arisen in cyberspace, and should not stifle this new world with inefficient regulation or censorship, it is foolish to say that the public has no sovereignty over what an errant citizen or fraudulent corporation does online. As the representative of the people and the guardian of democratic values, the state has the right and responsibility to help integrate cyberspace and conventional society.

Technology standards and privacy issues, for example, are too important to be entrusted to the marketplace alone. Competing software firms have little interest in preserving the open standards that are essential to a fully functioning interactive network. Markets encourage innovation, but they do not necessarily ensure the public interest.

4. Information is not knowledge.

All around us, information is moving faster and becoming cheaper to acquire, and the benefits are manifest. That said, the proliferation of data is also a serious challenge, requiring new measures of human discipline and scepticism. We must not confuse the thrill of acquiring or distributing information quickly with the more daunting task of converting it into knowledge and wisdom. Regardless of how advanced our computers become, we should never use them as a substitute for our own basic cognitive skills of awareness, perception, reasoning, and judgment.

5. Wiring the schools will not save them.

The problems with public schools – disparate funding, social promotion, bloated class size, crumbling infrastructure, lack of standards – have almost nothing to do with technology. Consequently, no amount of technology will lead to the educational revolution prophesied by President Clinton and others. The art of teaching cannot be replicated by computers, the Net, or by "distance learning." These tools can, of course, augment an already high-quality educational experience. But to rely on them as any sort of panacea would be a costly mistake.

6. Information wants to be protected.

It is true that cyberspace and other recent developments are challenging our copyright laws and frameworks for protecting intellectual property. The answer, is not to scrap existing statutes and principles. Instead, we must update old laws and interpretations so that information receives roughly the same protection it did in the context of old media. The goal is the same: to give authors sufficient control over their work so that they have an incentive to create while maintaining the right of the public to make fair use of that information. In neither context does information want "to be free." Rather, it needs to be protected.

7. The public owns the airwaves; the public should benefit from their use.

The recent digital spectrum giveaway to broadcasters underscores the corrupt and inefficient misuse of public resources in the arena of technology. The citizenry should benefit and profit from the use of public frequencies and should retain a portion of the spectrum for educational, cultural, and public access uses. We should demand more for private use of public property.

8. Understanding technology should be an essential component of global citizenship.
In a world driven by the flow of information, the interfaces – and the underlying code – that make information visible are becoming enormously powerful social forces. Understanding their strengths and limitations, and even participating in the creation of better tools, should be an important part of being an involved citizen. These tools affect our lives as much as laws do, and we should subject them to a similar democratic scrutiny.

Published online and in print, this declaration was signed by senior writers, contributors and editors of Wired magazine, by critics and authors, as well as by cultural pioneers such as Douglas Rushkoff, Andrew Shapiro and of course, David Shenk himself.[3]
Over the following years, the declaration was signed by over 2,500 people.[4] But then, as the millennium broke and a new era of Web 2.0 technologies arrived… it was all forgotten.

The internet wants to be free markets
Where the late 1990s had called for nuance, the mid-2000s

demanded a never before seen wave of unregulated optimism. While the markets remained shaken by the dot-com bubble, speculation for the internet's use as a social force reached peak frenzy with every new webpage representing a breeding ground for social revolution. Computers infiltrated classrooms, community spaces and homes, while communications technologies promised to transform the world into Marshal McLuhan's dream of a "Global Village".[5]

Over the years, the internet became synonymous with progress, democracy and freedom, an open space in which communication and collaboration could occur on a truly level playing field. The rise of citizen journalism snatched the power from elitist gatekeepers and redistributed the ownership of information to the masses. Unlike the legacy mediums of television, radio or print, the internet represented an opportunity for genuine two-way communication. Through blogs, webpages, vlogs and feeds, everyday citizens were finally given an opportunity to answer back – both to governments and to corporations alike.

As the mythology of the internet as a platform for freedom shifted from folklore to accepted fact, self-proclaimed cyber-activists sought ways to protect their unregulated utopia from the grubby hands of governments and profiteers. As the American author Stewart Brand famously said, "Information wants to be free"[6] – a comment that would one day go on to become the mantra of the cyberpunk, libertarian and even online piracy movements. For many technologists working in the early days of the internet, the idea that any money should be made from the web was the antithesis of their world view. The internet was about freedom of expression, collective output and, most importantly of all, expanding ideas for pleasure rather than for cash. Writing at the turn of the millennium, even a young Larry Page and Sergey Brin felt that their new Google search engine should never be corrupted through the inclusion of something as crass as paid advertising.[7] (How times have changed.)

The irony, of course, is that while technologists were fighting to stop corporatism and government intervention perverting the internet, they were creating exactly the kind of deregulated free market system in which corporatism is so easily nurtured. Armed with libertarian rhetoric and scared that government intervention would limit the internet's potential, the World Wide Web was allowed to evolve into a sort of Wild West, with fortune being found for both pioneers and bandits alike. What was once an experiment in creativity and individual expression now mirrors many of the issues and inequalities that have long since plagued the offline world. The Global Village has been replaced with a Global Mall. Predatory pricing strategies have helped switch out single-site ecommerce businesses for monolithic corporations. Even individual expression, which once represented itself in a plethora of blogs, forums and hideous GeoCities pages has now been switched out for the formulaic dropdown boxes of standardised Facebook profiles.

Everywhere you look online – both in the realms of the internet and mobile apps – you see monopolies, standardisation, and the very worst of predatory corporatism, all neatly rebranded as a new age of "digital disruption". The once questionable rhetoric of free market economics is now so deeply ingrained in the mythology of Silicon Valley that it is hard to remember that any other system of progress ever existed. The free market experiment[8] that failed so spectacularly in General Pinochet's Chile has found a new home online; a digital sandbox in which the concept of a wholly unregulated society can be beta tested before being "rolled out" into the offline world.

The fact that most of these technologies were paid for by government funding, military research and tax-backed academia is irrelevant. The accepted narrative states that free markets encourage the creation of new technologies, technology is progress, and all progress is good. That is the fundamental "truth" upon which our entire technoutopia is built.

The age of technoutopia

When planning a book, most writers don't decide what they intend to call their work until long after the process of writing is complete. For me however, the title of this book was obvious from the outset. Having developed my university dissertation around the theory of Media Ecology, I grew increasingly interested in the works of media critic Neil Postman, and in particular his classification of social history into different eras of technological obsession.

A humanist with a deep respect for both culture and the influence of religion, Postman viewed the notion of "progress" with a great deal of scepticism, worrying that the black and white, fact or fiction mentality of modern science provided limited room for moral interpretation. As Postman's works developed, it became clear that his scepticism of progress manifested itself most of all in a distrust for the inherent optimism that surrounds technology.

In 1985, Postman published *'Amusing Ourselves to Death'*, his first critique of television as a medium for cultural change. In it, he explains the need for television to convert all forms of knowledge and discourse into a visual spectacle,[9] perverting the content in order to better fit the device – or as Marshal McLuhan put it: changing the *message* in order to fit the *medium*.[10]

Following the release of *Amusing Ourselves to Death*, Postman worked with journalist Steve Powers to extend his thesis into broadcast news media. Writing in *How to Watch TV News*, the pair examined how the introduction of flashy graphics, narrative construction and commercial breaks all helped to turn factual news reporting into a new form of reality entertainment.[11] Once again, it was an example of how the *medium* inevitably perverted the *message*.

In 1992, Postman developed his seminal work *Technopoly*, which shifted from television to examine the impact of the new mass medium of the day – the computer. Here he considered how the notion of progress would once again change, and how, rather than reinterpreting everything as visual entertainment, the medium of computing would interpret everything as information to be processed.[12] To Postman this was the biggest threat of all, a machine that removed all opportunities for nuance, ignoring qualitative concepts such as culture and morality in favour of hard facts and raw data.

It was in this critique that Neil Postman defined his stages of technological development, breaking history down into *tool-using cultures*, *technocracies* and *technopolies*. The first of these stages has become near impossible to find anywhere in the world. Tool-using cultures only develop technologies for a raw functional purpose, whether to solve an urgent problem in physical life (the invention of mills, ploughs, spears) or to serve a symbolic or psychological goal (building castles, imposing towers, churches). As Postman explains, in a tool-using culture "technology is not seen as autonomous, and is subject to the jurisdiction of some binding social or religious system."[12]

In the second stage of development, the technocracy, tools play "a central role in the thought-world of culture."[12] Technology is no longer a tool, but a form of thinking in its own right; the development of new technologies becomes a part of culture, often replacing aspects of cultural life that came before. Within a technocracy, "tools are not integrated into the culture, they attack the culture. They bid to *become* the culture."[12]

The third of Neil Postman's stages is technopoly, an era in which technology no longer subordinates culture but has subsumed the role of culture altogether. Progress is the new education, data is the new knowledge, and consumerism is the new citizenry. As Postman explains:

> "Technocracy did not entirely destroy the traditions of the social and symbolic worlds. Technocracy subordinated these worlds – yes, even humiliated them – but it did not render them totally ineffectual.
>
> "Technopoly eliminates alternatives to itself in precisely the way Aldous Huxley outlined in Brave New World. It does not make them illegal. It does not make them immoral. It does not even make them unpopular. It makes them invisible and therefore irrelevant. And it does so by redefining what we mean by religion, by art, by family, by politics, by history, by truth, by privacy, by intelligence, so that our definitions fit its new requirements. Technopoly, in other words, is totalitarian technocracy."[12]

What Postman is describing here, whether intentionally or otherwise, is a free market system. A system in which laws do not restrict behaviour, yet certain avenues are closed off or destroyed by the mystical hand of market demand. We see this now more than ever before, with once vital notions such as privacy being quietly eroded by the "demands" of progress. Nobody has banned personal privacy in the internet age, there just simply isn't room for it. A once vital aspect of human existence and individual development has been removed, simply because the demand was not high enough for the market to sustain it. Now, in a move closer to George Orwell than Aldous Huxley, some technologists are even starting to argue that privacy never existed in the first place – a "fact" that increasing numbers are willing to accept.[13]

It is this level of technological domination that even Neil Postman failed to anticipate. Technology and the wider notion of "progress" are not only universally accepted as the heart and soul of our culture, they now have the ability to rewrite history; to blind people to reality. For the new generation, Silicon Valley entrepreneurs are the new celebrities, the new priests in an age of near faithful devotion to the idea of technology as an answer to all of life's ills.

So blessed are we with the miraculous technologies that surround us that we have lost the ability to even question whether they still represent genuine progress. At the same time, the very notion of progress has been clouded by the definitions of the internet age. Where once progress would have meant the advancement of life on earth, it has now come to mean something far more corporatist – the advancement of efficiency, productivity and production.

If one were to ask a stranger on the street whether the human race is progressing, they would probably agree and immediately point to technological advancements such as mass food production, international communication or simply just "the internet" as proof of our ongoing development. But at what expense do these technologies come? We live in an age of swelling populations, depleting resources and the imminent threat of anthropogenic climate change.[14] A mental health crisis sweeps the "connected" world, with online harassment, revenge porn and digital blackmail contributing to a thirty-year high in suicide rates.[15] Having promised greater interaction, a new generation of "herbivore" men have given up on physical relationships in exchange for the sexual variety offered by internet porn.[16] In care homes, elderly people are provided with Skype and "social robots",[17] offering their younger relatives an excuse to visit even less than they already do. Beneath the traditional internet, the dark web promotes weapons, drugs and sex trafficking. The digital high street has been run out of business by the exact same handful of tax-dodging monopolies that destroyed traditional retail. Spaces for public communication are more available than ever before, and yet these spaces are all privately owned. The conversations that happen within are moderated, mediated and harvested for advertising purposes. Our words, photographs and memories are no longer our own, but rather the intellectual property of search engines, social media sites and advertisers. To the new generation, however, this is all just progress. To question the decisions behind such developments would be to attempt to fight progress itself.

This is a new age of blind technological optimism. An age so enshrined in the Silicon Valley rhetoric of libertarianism that it has forgotten why the concept of governance existed in the first place. This is an age that extends beyond Neil Postman's vision for a technopoly and enters into a fourth era in the technological assault on culture. Here, technology is not only always "the answer", it has moved beyond the realms of questioning entirely. We have gone further than technological optimism, or even technological euphoria. This is the age of technological utopianism. We have moved beyond the *technopoly* and into a *technoutopia*.

Returning to realism

This book is an attempt to peel back the rhetoric of technoutopia, asking critical questions of both the internet and the connected technologies that surround it. The more that these technologies penetrate into our day-to-day lives, the more difficult it becomes to assess them in a fair and impartial way.

The arguments laid out within do not assume that technology is automatically a bad thing, but rather attempt to question the idea that all technological advancement is automatically a good thing. This approach is neither meant to be optimistic nor pessimistic, but rather a return to the technorealist logic that was lost after the millennium. It is not a call to unplug your router, delete your Facebook account or go and live in the woods, but rather an opportunity to reflect on the subtle ways in which technology can influence our lives. Most important of all, it is an examination of the assumptions that those of us living in a technoutopia have come to accept as fact.

Specifically, the structure of this book examines three key pieces of mythology that lie at the heart of modern technological discourse. These are:

1. That access to information is making the world more transparent and therefore more democratic.

2. That by limiting bureaucracy and cutting out middlemen the internet is providing day-to-day users with a more powerful voice and ultimately more control.

3. That by providing its users with constant access to information and communication, the internet is making us smarter, happier and more social than we have ever been before.

While internet technology has helped to enhance all of the above in very blatant ways, it has in many ways also led to their subtle subversion. This is not to downplay the achievements of modern technology, but rather to understand its effects and to encourage room for greater discourse around its use. In an age of 3D printed firearms and digital black markets we cannot go on claiming the same free market mantra of libertarianism.

Cyberspace does not exist. It is not some ethereal realm outside of our own. The internet is where we live, where we shop, where we communicate and where the next generation of children is destined to grow up. We should not stand for fear mongering, but we also cannot hide behind the rose-tinted glasses of technoutopianism. What we need is an honest and open debate around the introduction of new technologies into our lives. As Neil Postman once wrote, "when we admit a new technology into our culture, we must do so with our eyes wide open... No medium is excessively dangerous if its users understand what its dangers are."[9]

We must learn to view technology not as some autonomous force, but as the creation of fallible human beings. No line of code is without its own contextual bias. That is a fact that we as a society must address, not just in the corporate or government sphere, but in our own daily lives. We need to rethink our position as consumers and remember our responsibility as citizens, only then can we come to terms with what we lose whenever we gain something new.

We live in a finite world. When one thing is created, another must be destroyed. The march of new technologies is no different. Only once we understand this fact can we ensure that our definition of advancement is accurate and that our march towards progress is genuinely in the direction of a better world.

The internet encourages democracy, equality, and transparency

"A new social system starts, and seems delightfully free of the elitism and cliquishness of the existing systems. Then as the new system grows, problems of scale set in. Not everyone can participate in every conversation. Not everyone gets to be heard."

Clay Shirky

"Some people observing the media landscape today have wondered whether truth even matters anymore. Perhaps, they speculate, in the new information age reality is simply a matter of belief, not anything objective or verified."

Bill Kovach

Over the last ten years, the internet has become intrinsically tied with the democratic ideals of freedom, equality and transparency. Far more so than television, radio, or even the Gutenberg printing press, web technologies have been praised for bringing people together in open, honest communication at a truly equal level. While all of these technologies have in some sense or another helped to open the way to a more democratic society, each has been let down by either biased intermediaries or a naturally undemocratic infrastructure. In the instance of newspapers, the introduction of a critical press may have opened the public's eyes to corruption and hypocrisy in Whitehall, but any truly critical dissent still had to pass by the approval of millionaire media moguls and their profit-oriented editors. The same was true of the introduction of television. Despite the role that TV played in increasing transparency behind both sides of the Iron Curtain[1], the fact remained that very little could ever make it onto our airwaves unless it fell within the confines of general acceptability. The standards of this acceptability were once again defined by largely faceless mediators, working as either state representatives (in the case of public-owned programming), or as the corporate owners or sponsors of the television networks.

When compared to these legacy communication mediums, the internet has been widely celebrated as the first media to truly cut out such middlemen, minimising their negative effects upon the communication process.[2] For nothing more than the cost of an internet router, citizens all over the world can now communicate directly with one another unrestrained by editors, network operators or state-sponsored censorship. At least, that is the fairytale that those raised in a technoutopia like to tell themselves. According to utopian narratives, the internet has finally stripped communication of all bureaucrats, mediators and even the most basic forms of gatekeeper. Free of these biased middlemen, it is now an open medium, and – apparently – to be open, is to be free.

This narrative of freedom – and by extension, democracy – can now be found throughout every corner of internet debate, from academic literature through to corporate blogs. It is, to use the technoutopians' own favourite buzzword, ubiquitous.

One need only look at the works of popular internet theorists to see this obsessive, flag-waving, freedom-talk in action. The internet evangelist Clay Shirky claims that our social tools "remove older obstacles to public expression";[3] likewise, social media guru Brian Solis believes that we are "at the beginning of a new age of digital democracy".[4] Even more extreme is Jeff Jarvis' claims that we "don't need government in cyberspace" we only "need freedom".[5] I'd be interested to see the results of Jarvis' anarchist internet, as I'm sure would the owners of Silk Road and those hoping to download and 3D print unregistered firearms.[6]

But despite the widespread acceptance of this claim, the idea that the internet is a totally free space where members of every class, race or gender can talk without mediation is nothing more than a fantasy. In fact, as will be discussed later on in this book, the internet may well be the most highly-mediated communications device of all time.

For now, let us focus on the view that the internet is, as Shirky says, fostering "public expression".[3] According to the common narrative, web technologies are opening up dialogues between individual citizens and, as a result, are helping to make communication more transparent and ultimately more democratic. Already we have come across our first bare-faced assumption – that openness and transparency are intrinsically tied to democracy. Yes, the internet may well make communication more transparent (a claim that remains questionable), but to assume that this transparency will lead to a more democratic communication system is in itself an example of technoutopianism. The key flaws of this assumption were discussed at greater length by Russian political researcher Evgeny Morozov in his 2013 book, *To Save Everything Click Here*. According to Morozov, there are "right" varieties of transparency – which have positive impacts on democracy – as well as "wrong" varieties of transparency.[7] These wrong varieties may lead to "populism, thwart deliberation, and increase discrimination", all behaviours that could damage or stall the democratic process. To use Morozov's own example:

> "It is hard to believe that when Vladimir Putin orders workers to install webcams at polling stations across Russia, his invocation of transparency rhetoric serves

functions other than legitimizing his own stay in power."[7]

Yet despite such examples of undemocratic transparency, we in technoutopia continue to foster the idea that the internet must be democratic because it: *a.* encourages transparency, and *b.* provides users with an unmediated voice.

In addressing the first of these points, we have already seen that transparency represents little more than a tool to be wielded; it is not in itself an agent of democracy or positive change. As for the second, that the internet fosters democracy through a lack of mediation, it is growing increasingly clear that many of the now defunct gatekeepers – who the technoutopians so actively lament – were the only ones who could genuinely broker a balanced debate.

Without editors and network executives in place to review and filter the available content, the public sphere has become little more than an online shouting match, with opposing voicing left to squabble over the largest audience share. Having cast out the traditional gatekeepers, today's content producers have been forced to shout ever louder in an effort to be heard over the uncensored din of low-quality commentary. As a result, these producers have resorted to increasingly shocking – yet ultimately lower quality – content in an effort to push their products to the front of the crowd. This prioritising of shock over substance is what American author David Shenk once referred to as the "two-by-four effect";[8] suggesting that media producers must hit their audiences with a two-by-four just to grab a brief glimpse of their attention. As Shenk describes:

> "The two-by-four effect provides humanity with a way to keep communication alive in a glutted environment. But in so doing, it extracts a hefty price: Society, as we

all know from experience, is becoming inexorably more crass. We are witnessing the new reign of trash TV, hate radio, shock jocks, tort litigation, publicity stunts {and} excessively violent and sarcastic rhetoric." [8]

Faced with a near endless stream of media and competition online, content generators are being forced to lower the quality of their work in exchange for quick-hits and viral tactics.

At this point, it could be argued that such a phenomenon was never exclusive to the internet, and I would have to agree. If anything, the legacy of shock jock radio DJs and trashy gossip magazines is proof that all mediums have at some point succumb to the pressures of clicks over quality (or at least ratings over quality). The difference with older media however, was that - while low-quality works existed - they would always find their own private corner of the medium to inhabit and own. Celebrity gossip always belonged in glossy magazines. Likewise, hard news only really belonged in newspapers. Prank videos found their home on MTV, while political debate would rarely stray too far from the confines of BBC Radio 4. These boundaries did not blur. The idea of BBC Radio 4 conducting a pranks segment in an effort to expand ratings seems ludicrous. Similarly, the idea that a serious newspaper such as the Daily Telegraph or Independent would devote their pages to Take a Break style gossip seemed equally absurd.

And yet, in the unmediated internet age, these high and lowbrow pursuits are increasingly converged into a single space. Celebrity gossip is no longer the fodder of trashy magazines but rather a staple part of our online newspapers. Similarly, sites that once hosted serious political analysis are now just as likely to host viral spoof videos if they have even the weakest ties to politics (Musical political mashups being the most common example).

As further evidence, we need only look at the websites of the

Independent or the Daily Telegraph. Having followed their audiences online, both newspapers now find themselves battling it out to see who can write the gaudiest clickbait headlines. In the time spent writing this chapter alone, the Telegraph has run with "QUIZ: What does your favourite biscuit say about you?",[9] while the Independent has since opted for a more mature, "You can now ship your enemies a bag of dicks for them to eat".[10]

While it cannot be denied that this 'race to the bottom' mentality already existed within the offline world, it is also true that we have never come quite so close to actually reaching the bottom of that particular barrel – especially where the mainstream media is concerned. The question is, why have standards been able to drop so low?

As it turns out, this is what content looks like when you reduce the number of significant mediators. Before the internet did away with gatekeepers there was always some degree of editorial standards to ensure that the race to the bottom did not get out of hand. If a writer's efforts dropped for the sake of high ratings, you could be sure that a responsible editor would step in. If an editor failed to step in, sooner or later a senior executive would be likely to flag the story for bringing down the overall quality of the network. Why? Because prior to the days of internet clickbait, the development of an audience was widely considered a long-term partnership in which quality content was exchanged for loyal viewers. In the age of digital media however, limited attention spans and an abundance of choice have led viewers to flit between content, never quite settling on a preferred channel or site. This, added to by the near complete removal of mediators, has left content producers alone and largely unchecked. As a result, we are now seeing an age of media that is happy to provide misleading, factually inaccurate, and increasingly slanderous shock content in exchange for as many clicks as it can muster. This is the true beauty of "internet democracy", a brave new world in which all views are given equal, unverified, representation – regardless of whether they are true or not.

Now that we have abandoned traditional notions of quality, the only challenge left is to ensure that our content escapes the sea of clutter and successfully perches in front of as many eyeballs as possible. How? By turning up the shock and lowering the quality even further. This is the sad truth of the two-by-four effect, it can never really succeed and as such never truly ends. By demeaning articles with keyword stuffing and clickbait headlines, marketers and journalists are simply forcing their competitors to do the same. For every 100-word listicle written by The Independent, there will be a 90-word listicle written by The Daily Telegraph. There can be no winners in an online shouting contest, the result is a medium that just gets louder and louder until nobody is listening to anyone at all.

Paying for a megaphone

Despite the internet's democratic shouting match, not everyone online is being forced to dumb down their opinions in order to be heard. There are some people in the new technoutopia who are privileged enough to buy their very own megaphones, allowing them to shout even louder than the growing two-by-four din.

Yes, it turns out that the internet's democracy, in which everyone gets a voice, is in fact, a capitalist democracy. Instead of the heralded system of "one voice one vote", communicative democracy online is a system of "one dollar one vote", with the largest and most powerful players buying their way to the top of the communication chain.

Of course, offline democracy was never perfect in this regard.

Democratic equality has always been skewed by the need for campaign funding and the rise of cash-for-access lobbying and pressure groups. Similarly, the role of media in engaging and informing citizens has long been perverted by the deep pockets of advertisers and corporate lobbying firms. The question here however, is not whether democracy is flawed, but whether visions of the internet as a democratising and equalising force should be taken seriously.

If the internet is a genuinely separate space, then it should not be infected with the same corporatist biases of the offline world. Unfortunately however, the libertarian dream for internet freedom is instead increasingly resembling the freedom of a traditional market economy. As Australian economist Dan O'Connor explains:

> "Given that the government neither inhibits the activities of the internet nor props up or favors any particular actors or individuals, perhaps we are witnessing the closest thing to a free market that man has ever witnessed... a worldwide web that connects people from all parts of the world, allowing them to exchange whatever they want with one another. It is the essence of a free market: voluntary exchange."[11]

While there would be many economists who could argue that free markets are the pinnacle of democratic civilisation, past evidence suggests that once a market reaches a stage of truly unrestricted freedom, it grows increasingly likely that this freedom will start to erode the democratic infrastructure upon which it is based.[12] As Robert Alan Dahl, author of *On Democracy* says "Democracy and market-capitalism are locked in a persistent conflict in which each modifies and limits the other".[13] This conflict acts as a valuable tradeoff, in which markets lose some degree of freedom in exchange for the fact that individual "players" cannot exert undue influence that could undermine the democracy. Left entirely unregulated, these players can accumulate wealth and influence which puts them at a serious advantage over their fellow democratic citizens. In short, the voices of the few become more powerful than the voices of the many. Returning to the words of Robert Dahl:

> "Because of inequalities in political resources, some citizens gain significantly more influence than others over the government's policies, decisions and actions… Citizens are not political equals – far from it – and thus the moral foundation of democracy, political equality among citizens, is seriously violated."[13]

No matter how much internet technologists may triumph the idea that the internet will "fix" democracy, the truth is that web technologies are doing virtually nothing to solve this problem. In fact, given that the internet is "the closest thing to a free market that man has ever witnessed",[11] it is likely to make the problem worse. Just because it provides a structure in which anyone can have their voice "heard" (posted), that is not to say that powerful individuals and organisations will not be able to corrupt the system with undue influence.

As internet critic Andrew Keen notes, the internet is a "top-down system that is concentrating wealth instead of spreading it… distributed technology doesn't necessarily lead to distributed economics."[14] Those with the largest budgets will always be at an unfair advantage within the free market system. To have an equal voice is not to have an equal say, and this fact remains as true as ever in the internet age.

To expand upon this point further, let us take one particular example of internet communication that has been triumphed as an "equaliser" around the world – that of the humble internet blog.

According to many internet commentators, blogging represented the democratisation of the written word. No more publishing houses, magazine editors, or pesky qualifications required. If you could write, you could blog. What better example of an equalising technology that has thrown off the shackles of traditional publishing and banished those gatekeepers who held the industry back for so long. Pick up any technology textbook from the early 2000s and you can witness this rhetoric in action, usually making extensive reference to the comments of A. J. Liebling.

In the early 1960s, A. J. Liebling worked as a journalist for the American satirical publication, The New Yorker. In a now famous piece of commentary, Liebling claimed that "Freedom of the press is guaranteed only to those who own one"[15]. This pithy sentiment has long-since been quoted by technoutopians around the world, claiming that the introduction of free and intuitive blogging platforms such as Wordpress and Google's Blogger has finally laid such assertions to rest. Where once only the wealthiest in our society could afford to thrust their views upon the general population, now we can all "own" a private printing press in the form of a blog or a social media profile.

Out of all the technological revolutions that have evolved since the first days of the World Wide Web, this ability for everyday citizens to promote their opinions has been one of the most widely celebrated. From citizen journalism to Web 2.0, two-way communication lies at the very heart of our technoutopian society. Sadly, however, it too must only exist within the flawed confines of our free market internet economy.

As obvious a comment as it seems to be, the problem with providing everyone on the planet with an equal voice is that it starts to become awfully hard for anyone to get their point of view across. With every new voice that gets added to the internet, the overall influence of each individual commentator declines. As Clay Shirky notes in *Extreme Democracy*: "A new social system starts, and seems delightfully free of the elitism and cliquishness of the existing systems. Then as the new system grows, problems of scale set in. Not everyone can participate in every conversation. Not everyone gets to be heard. Some core group seems more connected than the rest of us".[16]

Returning to the example of the blogosphere, by January 2015, there was a total of 241 million blogs running on Tumblr and WordPress alone. This number was further added to by Google's Blogger, as well as lesser known platforms such as Typepad and Medium. Across all of these platforms, it is broadly estimated that an average of 68,000 blog posts are being created every sixty seconds. For every one of these new blogs that is added to the internet, the average "share of voice" provided to each commentator decreases.

Under such strenuously infobese conditions, even the most optimistic of internet pundits have been forced to admit that most blogs and posts will almost certainly never be viewed. To return to the comments of Clay Shirky, "in general… dozens of weblogs have an audience of a million or more, and millions have an audience of dozens or less".[17] Some technorealists are even more pessimistic, with Alexander Halavais – author of *Search Engine Society* – claiming that: "Of the millions of blogs in the blogosphere and videos on YouTube, most get viewed by only one or two people".[18] So much for owning your own printing press. At these circulation rates, most bloggers would be better off spouting their opinions at their local pub.

Perhaps that statement is a little unfair. After all, throughout much of the internet-centric literature, the 99.9% of blogs that remain essentially unvisited are actually considered a triumph of consumer choice. By providing internet users with variety, choice, and a broad spread of opinions, internet pundits claim that these "failed" blogs are equally important to online discourse as any Gawker, Politico or Huffington Post. According to their logic, the sheer availability of options is, in itself, evidence of the internet's inherent value and dedication to truth. As the argument goes, if you hold a strong opinion, you can always find its counter argument somewhere online. As a result, the 99.9% of blogs that are virtually unvisited are simply there to balance out mainstream consensus, providing a counter argument to help strengthen our knowledge of the world and ensure that we learn to question our own innate biases.

While this all sounds very useful and democratic, what such arguments fail to account for is the possibility that these blogs will not be relied upon to "cement truth", but will instead be used to confirm and reinforce existing beliefs. Quite opposed to knowledge or truth, the massive variety of ill-research opinions currently being shared online are instead providing a one-stop-shop for "evidence" on just about every unfounded assertion that could be fabricated in the human imagination. With the right Google search, internet pundits can reaffirm their views on anything from medical misinformation to theories of racial superiority. Given human beings' astounding susceptibility to confirmation bias, is it any wonder that the internet is so commonly used to justify existing incorrect beliefs? Whether it's in referencing academic articles, or simply "proving" a tall tale down the pub, the internet has become the go-to point of reference for supporting arguments without ever straying beyond the comfort zone of one's own opinion.

Yet despite this fact, we as a technoutopian society are still happy to believe that the internet is somehow linked to positive (if largely abstract) ideals such as freedom, consumer choice and the search for knowledge. Even when faced with the undeniable idea that most blogs will never be viewed – and most voices never heard – we continue to twist ourselves into believing an optimistic fabrication. To do anything else would be to undermine the very premise of the internet as a force for good.

We are the 0.01%

Not only are most blogs never viewed, the vast majority of them do not make it anywhere near page one of search results. Given that less than 5% of internet users search beyond the first page of Google,[19] most of the content being produced by the average internet users is essentially invisible. Furthermore, the 95% of content that resides beyond page 1 of Google is actually nothing more than the tip of a much larger iceberg, with a huge proportion of sites never even registering within Google's massive data web. For many sites, a lack of external references and poor SEO practices have made them impossible for search engine "spiders" to crawl. As a result, they remain floating in the unregistered ether now known as the "Deep Web".[18]

Coined in the early 2000s by computer scientist Mike Bergman, the term Deep Web is now commonly used to describe the massive collection of unseen web pages that never quite made it into Google's 30 trillion results.[20] As a result, the internet that users so regularly see and experience is nothing more than a sample of all the pages available online, with vast swaths of sites failing to meet the SEO criteria necessary to appear within most popular search engines.

While it is hard to comment on the exact size of the deep web – with most research into the topic having been conducted in the mid 2000s – its very existence points to the fact that most online content will never be viewed. Between the 99.9% of blogs and websites that will never make it onto page one of Google, and the massive amounts of content that will never make it onto

Google in the first place, most user-generated content online is never even seen.

While this fact can tell us a lot about the levels of empowerment attributed to blogging technologies, what is more relevant is that – by extension – the vast majority of internet users are being sent to the exact same handful of sites, over and over again. Instead of opening up millions of new channels of information to the public, it appears that the internet is falling into the same pattern as TV and radio, with most users accessing only a select handful of "channels" for their information, news and entertainment.

Faced with an over-abundance of choice, most consumers (whether online or offline) will turn to what American linguist and statistician George Zipf describes as the "power-law" distribution.[21] Applied to the online realm, this scenario predicts that the rapidly expanding quantity of web visitors will become increasingly reliant on an ever smaller population of websites. Similar examples of this distribution can be found throughout most consumer purchases, with research suggesting that as customers grow overwhelmed with options, they default to a small handful of trusted suppliers. Ironically, more choice may actually lead to poorer decision making. Philip Graves examined this phenomenon in depth in his 2010 book, *Consumer.ology*. Discussing the conflict between unconscious desires and rational thoughts, Graves considers the possibility that – despite what consumers may think they want – "choice" can, in fact, be perceived as a negative value:

> "When you ask them, most people say they want choice; often it will be a conscious consideration when selecting a retail outlet for a purchase – "I'll go to X because they have the biggest range." Choice is a good thing, isn't it? Social psychologists Iyengar and Lepper carried out an experiment that illustrated how, in practice, more choice

isn't necessarily beneficial. They evaluated reactions to two tasting tables at a supermarket; on one they laid out 24 different jams on the other just 6. While more people elected to stop for the wider selection (60% vs 40%), a dramatically higher proportion purchased from the selection of six jams, whereas only 3% did so for the larger choice. Put another way, less than 2% of people will buy from a display of 24 jams, but 12% will if you give them a choice of just six."[22]

In another example, closer to the topic of internet technologies, Google made a similar error when it surveyed customers to find out how many search results they wanted to see per page. As with Iyengar and Lepper's jams, consumers demanded more choice and as such Google tripled the number of search results per page. After witnessing a dramatic drop in web traffic, the search giant decided to ignore its users' advice and return the results to a mere 10 per page.[23]

When this principle is applied to the wider internet, is it any wonder that 99.9% of users are visiting 0.01% of sites? For all the rhetoric around empowerment and abundance of choice, it turns out that consumers do not actually want choice. What they want is a handful of powerful sites, providing them with easy options, quick decisions and comfortable familiarity. As Clay Shirky writes in *Here Comes Everybody*, "In the weblog world there are no authorities, only masses, and yet the accumulated weight of attention continues to create the kind of imbalances we associate with traditional media."[17] Once again, this does not seem to resonate too well with the view of blogging as a "great equaliser".

Still, even if the vast majority of blogs remain essentially unseen, this does not change the fact that a small handful have broken through into the mainstream media and now have their views shared with millions of people across the globe. For these lucky few, A. J. Liebling has been proved wrong. Thanks to the blogosphere, freedom of the press is not something that belongs exclusively to those who are rich or powerful enough to own a printing press – not anymore. At least, that is what we have to hope. Sadly however, any in-depth look at the 0.01% will quickly reveal that the distribution of power has barely changed, with the rich and powerful still dominating the online sphere, drawing much of their influence from their own offline empires.

Given that the internet is meant to provide everyone with an equal platform, it is interesting to note that all of the UK's top ten most visited websites are owned by either multinational corporations or government institutions. In fact, thanks to recent consolidation, most of the top ten sites are now owned by the same small handful of companies.

This issue is particularly obvious when it comes to news portals – places where people can go for the latest commentary and opinions to help them form their views. According to the manifesto of technoutopia, personal content generation should be smashing the monopoly of the mainstream media, replacing media moguls with citizen journalists capable of reporting the truth unmediated, direct from their smartphones. And yet, despite this vision, a glance at the most visited websites reveals that the traditional media are still the most active forces within the online sphere. While it may be true that anyone can now create their own printing press, the fact remains that promoting and distributing such content takes a huge amount of money, power and personal resource. Just because the upfront costs of producing content have been removed, that is not to say that the monopoly on information control has been altered. If anything, the costs of production have simply been moved further down the line, towards the advertising, promotion and distribution stage of the chain. As long as this remains the case, the internet will never succeed in separating itself from its offline rivals, nor in providing individuals with anything resembling an equal voice on the world stage.

More and more internet communication begins to resemble the offline "legacy" institutions that it was destined to destroy. From consolidation of ownership through to the introduction of intermediaries and gatekeepers, the pattern of internet development has provided an adequate parallel to the evolution of the newspaper industry almost 50 years ago.

Just as the number of US media owners shrunk from fifty corporations to six in less than thirty years,[24] the visible internet is little more than a handful of large businesses acting as gatekeepers to the world's information. While some of these corporations represent true entrepreneurship online (Facebook, Google, Twitter), in the vast majority of cases, these businesses are not even organisations that were spawned online, they are merely the old institutions dressed up as something new. Whether it's popular news blog The Huffington Post (owned by AOL), crowdsourced review site Rotten Tomatoes (owned by Rupert Murdoch's NewsCorp), or leading video game blog IGN (also owned by NewsCorp), the lines between democratic online media and corporate-sponsored offline media are already well and truly blurred.

At this point, it is probably worth reiterating the earlier statement that such an analysis may sound naive to those well-versed in the structures of economic capitalism. Obviously those organisations with the most money will have the best opportunity to invest in their ventures and, as a result, they will be the most likely to drive audience share and ultimately succeed. There is no business-centric conspiracy here. The likes of News Corp are not actively perverting the democracy of the internet; they are merely taking advantage of a system in which money and influence can naturally interchange. The point, however, is that the internet does nothing to resolve this issue. It does not provide a safe-haven outside of corporatist influence, nor does it provide a more equal or democratic soapbox upon which individual voices can be heard. The internet is simply one more medium for communication and transactions within a far wider social, economic and political ecosystem. Just because something has moved "online" does not mean that it will fix the commonplace problems of its offline alternatives. If the mainstream media is not providing individuals with an equal voice, no amount of blogging, webpages or tweets is going to change that fact.

The failure of journalism 2.0

As we have seen, despite the commonly accepted narrative that web technologies help to equalise individual voices within the public sphere, the truth is that the internet can be just as open to bias, monetary influence and PR spin-doctoring as virtually all legacy media that came before it. As Andrew Keen comments in *The Internet is Not the Answer*, "while all this technology might be novel, it hasn't transformed the role of either power or wealth in the world."[14]

Yet despite this fact, our utopian commitment to the internet has allowed us to ignore all of these downfalls, honing in on the technology's successes and imposing a selective deafness whenever news of its numerous failings reaches our ears. Like religious zealots, we worship our internet technologies as both omniscient and omnibenevolent – not only are they never wrong, but they can also do no wrong, only ever being used as a force for good.

It is this zealous oversimplification which once encouraged Facebook's Mark Zuckerberg to proclaim that the problems of the Middle East were not down to a "deep hatred of anyone", but were rather just the result of a "lack of connectedness and lack of communication".[25] Apparently, internet solutionism is now so powerfully ingrained within our collective psyche that the only possible solution we can find to some of the world's most serious political and religious problems is to "throw more internet" at them.

While the comments of Mark Zuckerberg may seem quaint, even humorous in their optimistic naivety, there is a darker side to such wide-eyed utopianism. At the end of the day, Mark Zuckerberg is not some quirky misinformed teenager, but rather the CEO of a $212 billion communications empire. The internal views and decisions of this empire not only impact how we communicate with one another on a day-to-day basis (through Facebook, WhatsApp and Instagram) but also how our communications infrastructure is managed and policed. Despite Silicon Valley's outwards facing claim that quirky tech firms do not "do" politics, the truth is that companies like Facebook are at their heart profit-oriented corporations that hold significant sway over the political process. Last year alone, Facebook spent more than $9.3 million on lobbying the US government, including hiring five top Public Affairs firms to help support their vast internal promotional machine.[26] When placed in this highly politicised context, Mark Zuckerberg's childish internet-centric opinions seem far more disconcerting.

And it's not just Silicon Valley that has been fooled into believing its own overly-optimistic rhetoric. In 2009, UK Prime Minister Gordon Brown declared that it would be impossible for a situation such as the genocide in Rwanda to ever take place again as, thanks to the internet, "information would come out far more quickly." According to Brown, "the public opinion would grow to the point where actions would need to be taken."[25] The fact that numerous wars and massacres have taken place since the popular adoption of the internet does not seem to enter into

Brown's thinking on this point. Instead, he chooses to ignore the specific details of *how* the internet will solve these problems and instead sticks with the upbeat narrative that it just will.

One thing this statement does make clear is that such 'internet-solves-all' opinions are no longer exclusive to Silicon Valley. Through years of inspirational speeches, public relations activities and government lobbying efforts, this overly simplistic mind-set has already started to seep into our political system. By repeating their sound bites over and over again – "the internet is freedom" – "technology is progress" – the technoutopians slowly convinced us that their explanation of the world simply must be true. As the politicians and social developers begin to clamber aboard, soon, ordinary people will start to believe them too.

Before you know it, internet technologies are offered up as the go-to solution for all our problems, even those that we did not perceive as problems in the first place. If we want to lose weight, we buy a Fitbit. If we want to get better at our jobs, we install a task management app. Need a solution to world poverty? How about a viral video? Looking to solve the Middle Eastern crisis? Just provide more communication channels like Facebook.

Ironically for a device that was meant to expand our horizons, in many ways the internet has left us trapped in a mental box in which the only solutions are those that involve either building or installing a new piece of connected technology. In many ways, this new-found solutionism is nothing more than an extension of the capitalist mentality, in which solutions to every problem can be found by making a purchase. The internet simply builds upon this view, covering it up with a shiny web interface.

Such solutionist mindsets are now so deeply ingrained within our society, both culturally and politically, that they are even beginning to warp the ways we act in the offline world.

Having seen how well the internet works – and having ignored those instances where it does not – technoutopians began to wonder why we were not applying the basic structures and ideals of the internet to all aspects of modern life. As they saw it, if the internet could make people more efficient, more equal and more democratically active online, then why could it not do the same for the offline environment? From this idea, spurious notions about information freedom, ad-funded content and unpaid crowdsourcing have started to spill over into the wider public domain.

After years of streaming TV shows, illegally pirating content, and reading the latest news online for free, is it any wonder that offline information has lost so much of its value? Now, when a consumer looks at a book, a video game or even a TV boxset, they no longer see the hours of work that went into their development, they simply see a block of information that, apparently, "wants to be free". And it's not just traditional entertainment formats that are suffering, even children's toys and games are succumbing to the effects of the online mentality. Following the launch of thousands of "free" (advertising sponsored) apps designed specifically for kids, many parents no longer expect to pay money for products to keep their children entertained. In a world where everything is translated into information and all information wants to be free, there is ever less room for any form of monetised incentive. Going forward, such an optimistic model must inevitably have unintended consequences for society. And yet, as thousands of businesses and entire industries struggle to meet the impossible new demands of the free economy, we continue to tell ourselves that it will all work out in the end. After all, it worked on the internet, so why should it not work out in the real world?

Nowhere is the potential damage of this internet-first mentality clearer than in the profession of journalism.

Having been convinced by their technoutopian colleagues that information must be free online, newspaper moguls placed their livelihoods in the hands of the internet gods and opened up their content online free of charge. As a result, print and broadcast journalists quickly found themselves in the impossible situation of trying to compete against their own content online. This has ultimately led to a rapid decline in physical newspaper sales – currently decreasing at around 7% per year.[27] While this may seem like an obvious eventuality for any business that starts offering its core product for free, the fact that media moguls allowed it to happen just goes to show how vested our faith has become in the power of the internet. Hard-nosed businessmen who had kept their ventures in tune with market demands for years, had opted to turn away from their basest instincts and offer their products for free, simply because "that's how things are done online". There was no long-term plan, no big strategy. They would simply start posting their content online, and – once again – the rest would surely work itself out. Sadly, however, it never did.

For those left to face the harsh realities of the offline, profit-oriented world, there seemed only one thing to do: restructure their failing news organisations to reflect what "works" on the internet. In an effort to compete with their online rivals, newspapers have been forced to continually lower their prices, ditching expensive investigative journalists, cutting salaries, and increasing their reliance on low-cost contributed content (often provided by spurious sources). By far the biggest impact of this cost-saving exercise has been upon the relationship between journalists and advertisers. As the average price of newspapers declines, editors have been increasingly forced to rely on subsidised funding from advertisers. This in turn, opened all the opportunities for bias and financially incentivised reporting that have come to plague journalism online.

While most respected journalists will deny that they would ever allow advertisers to dictate or influence their editorial content, as ad revenues grow increasingly vital to a media outlet's financial survival, it does not take a genius to see how easily this situation could change.

Even the likes of Larry Page and Sergey Brin once understood the detrimental impact sponsorship could have on independent content. Writing in a 1998 Stanford thesis, they claimed that Google would never rely on advertising for revenues as "advertising funded search engines will be inherently biased towards the advertisers and away from the needs of consumers".[28] If anything this should just be treated as common sense – although admittedly, Page and Brin's views on this subject have clearly altered somewhat in the last twenty years.

Regardless of what journalists say about the independence of their work, endless examples show that advertising does in fact influence editorial content. Many of these examples even show that such sponsorship bias was relatively common long before newspapers needed to compete with online alternatives.

In his seminal 1988 work, *Manufacturing Consent*, political philosopher Noam Chomsky unveiled his now famous five-step "propaganda model", which outlined how money and power can infect the independence of mainstream media, without the need for any global conspiracy.

Explaining the "filter" premise behind this model, Chomsky wrote:

> "A propaganda model focuses on this inequality of wealth and power and its multilevel effects on mass-media interests and choices. It traces the routes by which money and power are able to filter out the news fit to print, marginalize dissent, and allow the government and dominant private interests to get their messages across to the public. The essential ingredients of our

propaganda model, or set of news "filters," fall under the following headings: (I) the size, concentrated ownership, owner wealth, and profit orientation of the dominant mass-media firms; (2) advertising as the primary income source of the mass media; (3) the reliance of the media on information provided by government, business, and "experts" funded and approved by these primary sources and agents of power; (4) "flak" as a means of disciplining the media; and (5) "anti-communism" as a national religion and control mechanism. These elements interact with and reinforce one another. The raw material of news must pass through successive filters, leaving only the cleansed residue fit to print. They fix the premises of discourse and interpretation, and the definition of what is newsworthy in the first place, and they explain the basis and operations of what amount to propaganda campaigns. The elite domination of the media and marginalization of dissidents that results from the operation of these filters occurs so naturally that media news people, frequently operating with complete integrity and goodwill, are able to convince themselves that they choose and interpret the news "objectively" and on the basis of professional news values."[29]

Expanding upon the invisible nature of this system, in 1996 Chomsky was invited to discuss his propaganda model with BBC talk show host Andrew Marr. Perplexed by Chomsky's views of the media, Marr claimed that he was not "self-censoring" and that nobody had told him to ask the questions he was asking or say the things he was saying. As far as Marr was concerned, there was no puppet master whispering in his ear or pulling his strings. In response to this claim, Chomsky replied that "I'm sure you believe everything you're saying. But what I'm saying is, if you believed something different you wouldn't be sitting where you're sitting".[30] This simple yet damming quip provides a concise summary of the invisible nature of Chomsky's propaganda system.

While the basis of Chomsky's model is something that we will return to later on in this book, for now, it is simply being offered as a starting point for discussing the negative impact of sponsorship on the press. As soon as advertising becomes the main source of revenue for newspapers (whether online or offline), the nature of "news" is forced to change. To return to Marshall McLuhan's famous adage, "the medium is the message"; if the structure of the medium were to significantly change, then by extension, the contents of the message will also inevitably alter.

Switching to an advertising-funded model immediately changes this structure. As any businessman knows, the customer is the person paying the bills. Within an advertising-funded news structure, this person is no longer the *reader* but is instead the *advertiser*. As such, within most modern news organisations the advertiser has become the customer and the reader has become the product to be sold. Within this model, the actual news content of newspapers is nothing more than a way to entice the product (the reader), ready to be sold to advertisers in the form of clicks, circulation figures or average readership statistics.

In this quest to numerically justify advertising rates, modern newspapers have found themselves mimicking the business models of BuzzFeed and Gawker – increasingly relying on misleading headlines and scandalous photography in order to boost circulation figures. At the same time, the actual content of newspapers has grown increasingly simplistic and light-hearted, helping to reassure advertisers that the end product (the readers) will be in a suitable "buying mood". This positive mind frame is not something that advertisers are willing to leave up to chance, with numerous persuasive psychologists providing books and lectures helping marketers to place their audiences in a state of "cognitive ease". American psychologist Daniel Kahneman explained the benefits of this state within his bestselling book *Thinking, Fast and Slow*. According to Kahneman:

> "When you are in a state of cognitive ease, you are probably in a good mood, like what you see, believe what you hear, trust your intuitions and feel that the current situation is comfortably familiar… When you feel strained, you are more likely to be vigilant and suspicious: invest more effort in what you are doing, feel less comfortable, and make fewer errors."[31]

In short, if advertisers want their customers to be open to suggestion and uncritical of their advertising messages, they need to create an environment for their ads which provides a high level of cognitive ease. Unfortunately for newspaper publishers, nobody feels like purchasing garden furniture after reading a 2000-word opinion piece on the political history of Gaza. Instead, if newspapers can keep their readers in an aspirational mood, gawping over celebrities and reading reviews of fine wines, they will be much more likely to respond positively towards advertising. This, in turn, will make it much easier for newspapers to source advertisers who are willing to populate their pages. It will also stop those advertisers from having to worry about accidentally associating their brand with mass beheadings in the Middle East. It's a win-win for everyone involved… unless, of course, you're a reader looking for actual news.

Clearly, advertisers want to influence the content of news. However, once they are in a position to fund the media, not only do they have the power to impact the content of news, they also have the ability to influence which news sources thrive and which are allowed to disappear. As Noam Chomsky put it, "from the time of the introduction of press advertising... working-class and radical papers have been at a serious disadvantage. Their readers have tended to be of modest means, a factor that has always affected advertising interest."[29] As ever more papers and magazines switch to the internet-inspired view that information should be free, this bias towards affluent audiences is likely to become an increasingly significant counterweight to journalistic integrity.

While most mainstream journalists will dispute Chomsky's views in this area, the fact remains that, with or without the work of invisible filters, there are numerous examples where the press has directly bowed to advertising pressure and has changed its editorial content as a result.

One of the most famous instances of this came from the environmental organisation Greenpeace. In 1988, Greenpeace attempted to publish a piece of content in the Guardian, criticising the Ford Motor Company for lagging behind other car manufacturers from an environmental standpoint. According to the non-profit, Ford had failed to adapt its engines to take unleaded fuel without conversion and had opposed the fitting of environmentally-friendly catalytic converters throughout the European market. In an unabashed instance of sponsorship bias, the Guardian refused to publish the story for fear that the Ford Motor Company – then one of the largest advertisers in the UK[32] – would be deterred from placing ads in future editions of the paper.[33] Clearly, even before the advent of online news, journalists were still feeling the pressure to conform to the demands of large advertisers. Now, with newspaper editors growing increasingly reliant on advertisers to keep their publications afloat, such blatant disregards for editorial standards are only going to get worse.

Sadly, the behaviour of the Guardian does not seem to be a one-off mistake. In 2010 an advert for Amnesty International was pulled from the FT as it ran the risk of offending Shell – a key advertiser for the FT's parent company, the Pearson publishing group.[34] While many journalists still advocate that such examples are nothing more than one off mistakes, these mistakes do seem to be occurring on an increasingly frequent basis.

In early 2015, BuzzFeed editor Ben Smith admitted that some articles had been previously removed from the site as they were considered critical of key advertising partners.[35] In one instance, BuzzFeed staff facetiously referred to Unilever's "Lynx Effect" advertising campaign as advocating "worldwide mass rape." Following an internal complaint from Unilever – a core advertiser on the BuzzFeed network – the article was quickly taken down from the site. While the accusation of a brand advocating rape is obviously very serious (even if it was meant in jest), the idea that Unilever was provided with special treatment because of its sponsorship status is concerning to say the least. Here we have a cut and dry example of a mainstream media outlet, which provides news and opinions to over 200 million people,[36] directly censoring its content in order to keep a large corporation happy. And yet, BuzzFeed is still widely considered one of the great success stories of internet journalism, with older outlets now basing their content strategy on the BuzzFeed site. Somehow, thanks to the imposition of internet-centric values upon the press, the idea of a publication censoring editorial content for cash is considered so unexceptional that it doesn't even merit any outrage.

Given the years that Silicon Valley optimists have spent building the links between technology and democracy, it seems ironic that internet technologies appear to have so effectively destroyed one of democracy's core building blocks – freedom of the press. By swiping aside quaint "legacy" values such as paying for information, the internet has helped lay the foundations for an entirely new approach to journalism. Journalism 2.0; a new press for the new era of technoutopianism.

Thanks to the work of existing journalists, the basic infrastructure of this new journalism is already well under way. By making content available for free, we have successfully developed a model of reporting that allows people the freedom to express their views, but only if that expression will help to generate profits from advertisers. This is a press that does not sell news to the public, but rather sells the public to corporate sponsors. It is a press that, while reluctant to support censorship, is increasingly forced to censor itself in order to stay afloat. Often, it is a press that lies to its readers; offering incorrect headlines and irrelevant imagery to increase clicks and circulation figures. In many ways it is a press that writes purely to entertain; avoiding gritty or complex subjects in order to keep audiences in a malleable "buying mood". It is a press that does not "do" politics. A press that cannot afford investigative journalism, and certainly cannot afford to write articles of any depth or quality. This is the press of a technoutopia.

The success of spin 2.0

While Journalism 2.0 may have failed in its efforts to conquer the internet and survive without significant deformity, the new age of public relations has proved a far more successful venture. Whereas every new communication channel opened by the internet represented a potential threat to existing journalism, the very same channels offered a whole new wave of opportunities within which PR professionals could ply their trade.

Before going into the details of how and why PR has proved so successful online, it's worth taking a bit of time to reiterate what

"PR" really means, and what exactly it is that the public relations industry does.

Despite its significant and growing role in our society, for many people, public relations remains a mysterious non-entity. As far as most members of the public are concerned, newspapers, magazines and TV news are all still seen as the exclusive outputs of professional journalists. In reality, this has never really been the case.

As it stands, around 55% of newspaper content is developed, not by journalists, but by members of the public relations industry.[37] This figure jumps even higher for glossy magazines.

Even broadcast news content increasingly relies on pre-recorded video and audio segments paid for and packaged up by corporations. These "video news releases" (VNRs) are used by companies to either subtly promote a product or to maintain some degree of control over how an organisation's industry is portrayed in the press.

Given that the existence of an independent press is so fundamentally tied to democracy, it is astounding to see how few people are even aware that the public relations industry exists – let alone that the majority of their daily news is provided by corporations. This lack of awareness on behalf of the general public is in no way accidental. In contrast to advertising and marketing, PR almost always benefits from working purely behind the scenes, becoming more persuasive by developing promotions entirely through the mouthpiece of third party endorsers.[38] The author Sylvia H Simmons summed up this difference with the example of a first date:

> "If a young man tells his date how handsome, smart and successful he is – that's advertising. If the young man tells his date she's intelligent, looks lovely, and is a great conversationalist, he's saying the right things to the right

person and that's marketing. If someone else tells the young woman how handsome, smart and successful her date is – that's PR." [39]

This is the role of public relations, remaining invisible while third parties such as journalists, celebrities, scientists and even politicians endorse brands on their behalf. In order for such endorsements to prove beneficial, the hand of public relations must stay permanently out of sight. In this sense, the use of PR in marketing is much the same as the use of placebos in medicine. If a patient is informed that they are taking a placebo, then the illusion is shattered and the placebo effect does not add any benefit. Likewise, when a member of the public is informed that a third-party endorsement is not genuine, or that the endorser has some vested interest in what they are endorsing, then the benefits of PR will be immediately lost. As a result, public relations professionals have a deeply vested interest in ensuring that their role remains either unseen or misunderstood by the public at large.

This widespread misunderstanding is not only the work of the public relations industry, it is also encouraged by journalists themselves. While there remains a dislike for PR professionals throughout the world of journalism, the truth is that most journalists rely heavily on PR to make their working lives significantly easier. Faced with tight deadlines and low research budgets, many journalists are more than happy to accept pre-written contributions from PR firms in order to fill space and generate regular content. Similarly, for journalists looking to speak with celebrities or high-profile corporate figures, public relations provides a helpful – if stringently controlled – portal of access. Given that so many journalists now rely on public relations content to help develop their stories, efforts to criticse the PR industry can prove equally detrimental to the journalists themselves. In this sense, keeping the extensive role of PR quiet is necessary for both industries.

The one area where PR has received wider attention is within entertainment, with television shows such as *Absolutely Fabulous* and *Sex and the City* often focusing in on the complex lives of glamorous PRs. While such shows have at least helped to bring public relations into the mainstream vocabulary, they have also resulted in an extremely skewed view of what it is that the PR industry does. According to these portrayals, public relations is always one of two things: either glamorous girls meeting up with celebrities, going to parties and drinking champagne (*Absolutely Fabulous, Sex and the City, Twenty Twelve*), or, as Machiavellian spin doctors, sitting in dark rooms and pulling the strings of politicians and the press (*The Thick Of It, Thank You For Smoking, Man Bites Dog*). While both of these elements do occasionally fall into the remit of public relations, the vast majority of what PR professionals do is far more mundane and – worryingly – more institutionalised within our current media environment.

While the field of PR still lacks a formal definition, traditionally speaking the role of public relations is to facilitate understanding between a business and its various stakeholders. These stakeholders could be potential customers, members of the press, analysts, employees, environmental groups or politicians – essentially, anyone who has an interest in the business or could potentially have a detrimental effect on an organisation's bottom line. According to the *Chartered Institute of Public Relations*, an official body for the industry, it is then the role of public relations to open up a "two-way conversation" with these groups, helping them to appreciate the ways in which their goals and the organisation's goals could potentially overlap.[40] What PR professionals are supposed to do when these goals do not overlap is not specified. Although, as someone who works in the public relations industry myself, I think it's fair to say that the client (the corporation) often comes out on top.

While targeted stakeholder engagement can prove an extremely vital business asset, sadly, many people in the PR industry never actually attempt such high-level strategic thinking. Driven both by client demands and a need for tangible results, most PR activities have become synonymous with "media relations" – a specific field of PR focused solely on building coverage in the press. This coverage has widely become seen as an end in itself; typically secured by arranging interview opportunities, distributing press releases, and developing articles for placement in targeted magazines. In short, modern PR professionals spend most of their time trying to keep their clients' names and website addresses in the media.

As the PR industry has grown increasingly comfortable defining itself in these terms, the various other aspects of stakeholder engagement have either been split off or sucked up by other overlapping disciplines. Managing a brand's customers has become the duty of marketing departments and dedicated customer service teams. Likewise, managing reputation in the eyes of potential investors has become the responsibility of CFOs and financial advisors. Even lobbying – once a staple of the Machiavellian PR role – has been annexed off to form its own external discipline in the form of "public affairs".

At this point, public relations probably sounds like a fairly toothless animal, focused more on celebrity gossip than anti-democratic Machiavellian antics. And yet, as a trade it remains a continuous source of controversy. Books with provocative titles such as *"Dark Art", "Trust Me, I'm Lying" and "Toxic Sludge Is Good For You"* continue to hit the shelves, providing a detailed and devastating history of the unethical, mischievous and outright undemocratic things that public relations professionals have achieved in the past. While it is true that public relations now gets away with less than it used to, it's difficult to deny that as an

industry, the fundamental activities of PR are hardly democratic.
One argument used within the field is that working in PR is no different from being a defence lawyer. In the same sense that everyone deserves some form of representation within a court of law, PR professionals would argue that every company deserves representation within the court of public opinion.[41] The difference, however, is that when one person is found innocent in a court of law, another is not automatically sentenced in their stead. In the cutthroat market environment however, for every business that gains market share through expensive PR activities, another is forced to lose out. This creates a system in which those companies that can afford the highest levels of PR investment will typically benefit from the best publicity.

While this may sound like an obvious outcome of monetary success (more profit means more opportunities for investment and growth), the problem comes when a firm offers worse products or services than its competitors but can still afford far superior promotions, exposure and "spin" within the public domain. As someone who works in the public relations industry, I can say first hand that I have previously helped to promote a few products that were categorically worse than their competitors. And yet, rather than doing what was best for the consumer – or even best for society as a whole – PR professionals must continue to represent those clients that will pay the largest sums. No matter what sympathetic bodies such as the CIPR say about fostering "two-way communication", at the end of the day, no PR agency is going to refuse business because a competitor's product would be better for society. More often than not, public relations is about providing publicity for those companies that are willing to pay for it. Those with the most money will inevitably get the best PR.

As we saw in the previous section, journalism is increasingly under siege from the new era of free information and internet technologies. Faced with lower budgets, shorter deadlines, and ever greater demand for real-time content, internet journalism has become the perfect target for the PR industry. In order to keep churning out regular content online, journalists have become ever more reliant on announcements and press releases as inspiration for their stories. Likewise, in a bid for interesting content, journalists are also now far more open to the idea of accepting entirely pre-written content in the form of "bylined" articles. Such articles are typically positioned as contributed opinion pieces, providing the topical views of a celebrity, spokesperson, or even the head of a corporation. If you have even seen an article in a magazine supposedly written by someone like Richard Branson, talking about a topic such as the future of space travel, this is probably a bylined article. In the vast majority of cases, such articles have never even been read by the supposed author and are instead the sole creation of a public relations agency – presumably looking to raise awareness for Branson's new space tourism venture. More often than not, even the magazine editors themselves are in on the ruse. Journalists are not stupid. They are fully aware that the articles provided to them are nothing more than fabrications, but in the internet age such falsehood is seen as little more than part of the process of filling space, gaining clicks and selling adverts.

Throughout my own career in PR, I have often seen the extent of such editorial complicity. Several years ago, I was required to "sell in" a series of interviews with the CEO of a leading drinks company. The process of selling in has become a staple of the modern PR approach, but is more often than not synonymous with cold calling customers within a typical sales role. PR professionals prepare a media opportunity and then ring journalists selling in the idea and asking them if they would like to use the suggested content within their publications. In the case of interviews, if a journalist accepts the invitation, a PR person will then set up a call or a meeting date upon which the interview will take place. Since the advent of digital journalism it has become increasingly common for journalists to simply provide a list of "Qs" (questions) and then demand a list of pre-written "As" (answers) for the interview. Once again, the journalist is fully aware that the "As" will be written by the PR agency but, within the desperate confines of the attention economy, such an approach proves a helpful alternative to the time-consuming interview process.

While this may seem like a harmless enough practice, by circumventing the interview process journalists are allowing corporate figures (or more accurately their PR agencies) to skew the conversation without any additional questioning or probing on the part of the journalist. This removes a fundamental investigative element from one of the oldest aspects of the journalistic profession. Instead, businesses are left with an open invitation for blatant, unaccountable propaganda (the original term for PR). In one of the worst instances of journalistic ineptitude that I've ever personally witnessed, not only did a journalist ask me to draft the answers for his interview, he also asked me to draft the questions. This meant that I (by which I mean the corporation) had complete editorial control over the entire process. This is a slippery slope for the modern press to be traveling down.

Of course, if a PR agency's content is considered overly promotional then a journalist still maintains the right to refuse it. However, as journalists grow increasingly desperate for content and the standards of quality decrease online, PR professionals continue to push the limits of what counts as news and what is pure promotion. As a result, in most instances, once an interview has been emailed through, increasing numbers of journalists will simply proofread and upload it to their sites.

But it is not just written materials that are being manipulated by

PR professionals online. As improved technologies have helped to drive down the cost of video editing and production software, public relations has expanded to include not only the content that we read online, but also the content that we watch.

While broadcast media has never been quite as susceptible to the public relations industry as print, it has not managed to escape entirely. Since the 1980s, many of the larger PR firms have offered broadcast services that will help their clients gain coverage on television news and radio broadcasts. While the most innocent of these tactics simply involve "selling in" interviews with leading spokespeople, there are a number of additional methods that remain largely outside of the public's view. One of the most controversial of these tactics is the production of Video News Releases (VNRs). As with traditional press releases, a VNR will include details of a particular story along with pre-filmed video footage that broadcasters can include within their shows. While most of this footage is provided as unedited "B-roll", there are occasions when an entire segment will be provided that journalists are simply expected to drop into a news broadcast almost entirely unedited.[41]

While such tactics have long been frowned upon, with many communications theorists arguing that the source of such contributed material should be clearly labelled, the internet has once again shifted the balance of power away from journalists and into the hands of PR.

Another area where PR has increasing influence is in ensuring publications receive appropriate positioning in search results. In the fight for audience share, search engine optimisation (SEO) has become a vital tool for all journalists and publishers. As far as most internet users are concerned, if an article or publication cannot be found on the first page of Google's search results, then it may as well not exist at all. In fact, according to research by the search advertising company Chitika, only 4.9% of internet users ever make it to page 2 of Google. This figure drops to a mere 1.1% for page three.[42] Based on this fact, journalists are more conscious of SEO than ever before.

As a result, when research from Forrester revealed that web pages including videos are 53 times more likely to rank on the first page of Google,[43] the race was on for journalists to find as much video content as they possibly could. Unfortunately, nobody seemed to realise that this was yet another race to the bottom of the journalistic barrel.

As editors began to increase the pressure on their staff to include video content, journalists became increasingly aware that the production of high-quality video does not come cheap. Thankfully, PR professionals were once again waiting in the wings, eager to provide journalists with the content they needed to stay on top of Google's search results. With often extortionate marketing budgets to spend (compared to the declining budgets of internet journalists) PR agencies were more than happy to dip their hands into their pockets and start developing videos to supply to the press. Through the introduction of low-cost video hosting services such as YouTube and Vimeo, PR professionals were even able to provide a handy HTML embed code, which allowed journalists to simply drag and drop videos directly onto their sites.

As it stands, virtually all online newspapers, internet magazines and digital trade publications regularly embed video content alongside their articles. While sites such as the Guardian and Telegraph avoid the drag-and-drop mentality of most internet journalists by relying on their own video players, they are still far more susceptible to packaged content than their traditional broadcast alternatives.

Once again, by increasing the pressure on publishers and removing many of the barriers to entry, the internet has successfully undermined traditional journalism in favour of corporate PR.

But it is not just video content that PR professionals have accosted in their efforts for promotion in the internet age. It turns out that public relations is extremely versatile, coming up with hundreds of creative ways to manipulate the internet to the advantage of corporate clients. Some of the most interesting techniques used online have been outlined by Ryan Holiday in his 250-page exposé of the PR industry, *Trust Me I'm Lying*. As a former spin-doctor for American Apparel, as well as an editor for the New York Observer, Holiday has certainly done his fair share of internet manipulation. Many of the strategies that Holiday describes are not even considered traditional PR techniques, they have instead been developed purely for the internet environment. As such, rather than the bad habits of a hundred-year-old industry seeping into the online world, it appears that the shaky journalistic values of the internet are creating entirely new forms of propaganda online. This from a medium that is supposedly making the world a more transparent, equal and democratic place.

One of the most fascinating techniques of digital public relations is a process described by Holiday as "trading up the chain". This is used to manipulate the pre-existing biases of internet journalism in order to turn relatively unimportant and uninteresting announcements into national, and even potentially international news stories.

Holiday outlines the process of "trading up the chain" as follows:

> "It's a strategy I developed that manipulates the media through recursion. I can turn nothing into something by placing a story with a small blog that has very low standards, which then becomes the source for a story by a larger blog, and that, in turn, for a story by larger media outlets. I create, to use the words of one media

scholar, a "self-reinforcing news wave". People like me do this every day."[44]

While I would argue that this process is not quite as easy as Holiday makes out, trading up the chain has become an increasingly significant tool in the arsenal of online public relations. As the internet shortens deadlines and increases the demand for real-time content, journalists grow ever more likely to rely on copy-and-paste material from other smaller sources. In one study conducted by George Washington University, as much as 89% of journalists have reported using blogs as the source material for their news content.[45] Such findings provide an even greater incentive for PR professionals to follow in the footsteps of Ryan Holiday's manipulative techniques. Having spent years trying to influence the pages of national newspapers only to be thwarted by aggressive editors, lazy research techniques – encouraged by the internet – have finally allowed PR professionals to bypass such hurdles. Instead of attempting to convince journalists directly, in the internet age it is far easier to simply manipulate the source material from which they generate their news. By moving the practice of PR further upstream, public relations professionals are literally using third parties to influence their own third parties.

The profession of journalism is being torn apart, both from the outside and from within. The role of internet technologies in supporting this demise cannot be overstated. And yet, in the face of all this evidence, we are once again told that the internet will somehow produce a miraculous solution…

The internet will find a way

As we have seen, the internet has successfully managed to drown out minority viewpoints, build an oligopoly on information, erode the values of effective journalism, and ultimately replace a once independent media with an amalgamation of PR, clickbait articles and corporate sponsored "advertorials". And yet, as a society, we do not appear ready to even admit that such failings exist, let alone that they could prove a long-term flaw in a much revered system. As far as most are concerned, the internet has provided society with low-cost goods, flexible working, streaming videos, one-click purchasing, Uber, eBay, Amazon, TaskRabbit and Tinder, not to mention instant video and voice access to friends and loved ones. As a result of these seemingly endless "free" gifts, the internet remains somehow outside of the realms of public scrutiny – even when such scrutiny is required to improve the system or society as a whole.

When faced with the numerous, and often qualitative complexities created by the internet, the loyal residents of technoutopia turn once again to their one and only messianic solution: the internet. Stuck in an infinite feedback loop, the pundits of Silicon Valley are so blinded by their creation that they physically cannot perceive a problem to which the solution does not involve the development of yet another connected technology. Even if such problems were caused by connected technologies themselves; apps, websites and networked hardware will forever be the solution.

Ironically, in many ways these zealots have committed the cardinal sin of the public relations industry – they have fallen for their own spin. Somewhere between lobbying lawmakers on the democratic nature of their technologies and schmoozing the public with talk of transparency and social responsibility, the technoutopians have become convinced by their own hyperbole. Now, when Mark Zuckerberg claims that the intricacies of geopolitical warfare can be solved through a higher degree of "connectedness",[46] he is, in fact, being entirely serious. The once spurious notions that improved access automatically means improved democracy, that more information automatically means more transparency, and that greater efficiency automatically means a greater society, have become so fundamentally ingrained within all discussion of technology that they are beyond question. We are so invested in the internet's flamboyant myth that the mundane truth seems positively nonsensical. No matter how complex the problems, the internet *must* be the solution.

As a result of this unwavering commitment, when faced with a growing number of cases where the internet has led to less equality, less transparency and less democracy, the pundits of technoutopia once again turn to the only solution they know. As such, when addressing the decline of print journalism, the hopeful defenders of Silicon Valley informed us that internet technologies would provide an alternative, with citizen journalism pegged as the ideal replacement for the outdated field. By cutting out the corporate middleman, mobile cameras and blogs were going to put the power back into the hands of "real" people. This collective approach was destined to not only provide a more efficient media system, but also to ensure a more open and transparent press. As journalists Bill Kovach and Tom Rosenstiel write in their 2009 book *Blur*, the internet was supposed to herald "a citizen media culture that instantly self-corrects – a kind of pure information democracy".[47] Sadly, however, this experiment in crowdfunded news content has proved a colossal failure.

By taking reporting out of the hands of accountable professionals, citizen journalism has allowed misinformation to run riot. As Steve Waddington put it in his 2012 book *Brand Anarchy*, "for every example of citizen journalism that has broken a story, there are countless examples of social media being used to push propaganda or news stories that are plainly incorrect."[48]

One of the most infamous examples of such misinformation was seen during the Boston Bombings terrorist attack in April 2013. Following the detonation of two pressure cooker bombs during the final stages of the annual Boston marathon, police and federal investigators were left with few leads in their hunt for the perpetrators. Faced with limited alternatives, FBI officials did something that had never been done before; they turned to the power of social media to solve the case.[49] With more than five thousand people in attendance at the marathon, police felt sure that somebody somewhere must have either filmed, photographed or maybe even tweeted the perpetrators. In what was later described as "one of the biggest online crowdsourced investigations ever",[50] FBI officials took to Facebook, Twitter and Reddit to ask the public if they had seen or captured anything suspicion during the event. The local police force even launched its own dedicated website which allowed users to submit their tips and witness statements directly to the FBI.

But it was not just the authorities who turned to social media during this crisis; having realised that the police were no closer to identifying the culprits, the mainstream media also set their sights on Twitter and Reddit to help generate leads. This not only helped broadcasters fulfil their desperate desire to appear relevant in the digital age, but also helped to maintain audience interest in an otherwise blank 24-hour rolling news schedule.

With both the media and police having placed their faith in the crowdsourced justice of social media (some might say "vigilantism"), the technoutopians' hypothesis for an internet citizenry were finally put to the test. Sadly, however, they inevitably failed to deliver.

Relying on a combination of hearsay, speculation and dodgy photoshop, Redditors (members of the social forum Reddit) identified the culprit as Sunil Tripathi, a Brown University student who had been missing since March 2013. As the internet mob jumped on this unfounded speculation, Sunil's family was subjected to a barrage of threats, questions and abuse across multiple social networking platforms. As the news spread that Twitter and Reddit had identified a potential suspect, US talk radio host Greg Hughes tweeted: "Journalism students take note: tonight, the best reporting was crowdsourced, digital and done by bystanders." [51]

Of course, as we all now know, the missing Sunil was in fact innocent of the crimes put to him by the unchained internet mob. Eight days after the bombing – when Sunil's missing body was found floating in the Providence River – his bereaved and abused family were finally left in peace. Now almost four years after his false accusation, a Google Image search for this murder victim's name still ties him to one of the worst terrorist attacks of the last ten years.

Having failed to learn from the accusation of Sunil, Reddit next turned its investigative journalism skills against Salah Barhoum, a 17-year-old Massachusetts boy who happened to be photographed carrying a large sports bag at the time of the bombing. Once again, in an effort to copy the "real-time" (inaccurate) news generated by social media, the New York Post decided to include images of this individual on its front page, claiming that he was being sought by federal investigators.[52] Having spotted the headline, Salah quickly contacted police to clear his name. As with Sunil Tripathi however, he continued to receive abuse on social platforms for days after the initial accusation. This ultimately led him to deactivate his Facebook account five days after the bombings.

Following these examples of failed social media justice, the FBI was forced to release the following statement:

> "Contrary to widespread reporting, no arrest has been made in connection with the Boston Marathon attack.

> Over the past day and a half, there have been a number of press reports based on information from unofficial sources that have been inaccurate. Since these stories often have unintended consequences, we ask the media, particularly at this early stage of the investigation, to exercise caution and attempt to verify information through appropriate official channels before reporting."[53]

In short, the media needed to stop mimicking the views of an uninformed internet mob and start acting like professionals. It is interesting to note however that even the police could not quite bring themselves to damn the behaviour of social media itself. Rather than choosing to criticise the internet for its widespread promotion of disinformation, official sources chose instead to blame legacy media for not checking their facts with enough rigorous scrutiny.[54] Fortunately, this should not prove a problem for future generations, as it may be unlikely that such legacy media will even exist. The internet is a replacement technology, not a supplement. If traditional media continues to decline at its current rate, the reactionary, unchecked citizen journalism witnessed through the Boston Bombings will be the only form of journalism left. In some ways Greg Hughes was right – journalism students should take note of the Boston Bombings. Not however, because the best reporting was "crowdsourced", "digital" or "done by bystanders", but because this is the sort of disastrous internet speculation that they should spend their careers learning to filter and fight against.

And yet, despite the dark roads that such mob journalism has taken us down, our faith in the internet and our tendency to crowdsource all of society's problems is stronger than ever before. Even after all the speculation and false accusations, the Boston Bombings are still held up by many internet pundits as a story of social media success. Commenting shortly after the capture of the actual perpetrators, Wired magazine's Spencer Ackerman wrote of the event:

> "Hiding in plain sight was an ocean of data, from torrents of photography to cell-tower information to locals' memories, waiting to be exploited... the public's eyes and ears are crucial investigative assets, as the internet rapidly compressed the time it took for tips to arrive and get analysed".[55]

This pro voluntary-surveillance mentality ultimately led Ackerman to conclude that "in an earlier era, law enforcement might not have identified the suspects in the Boston Marathon bombing so rapidly". [55]
Once again, by ignoring the numerous cases of false identity, the attacks on innocent bystanders, and the abuse of bereaved families, technoutopians are able to reassure themselves that – thanks to the internet – things will always work out in the end. Somehow, through a process of wishful thinking and collective memory loss, the history of these events is rewritten. The stories of Sunil Tripathi and Salah Barhoum are now considered nothing more than natural hiccups in the early development of social media, a healthy teething phase in the organic process of crowdsourced journalism.

But it is not just citizen journalism that is expected to plug the sinkhole of professional media and reporting. Well aware of the problems currently being caused by click-hungry journalism and dishonest PR, many internet pundits are now looking to develop technological fixes that will help sniff out disinformation and strengthen the internet's faltering claim to openness, transparency and democratic equality.

In an effort to address the online media's growing reliance on corporate press releases and copy-paste "churnalism", Google used its January 2014 "Panda" update to introduce automatic penalisation of those sites that relied heavily on stolen or duplicated content. Through this update, Google ensured that any website that was found to contain content identical to that published elsewhere on the web, would be marked as low-quality and automatically pushed down in the search rankings.[56] As we have already seen, faced with reduced budgets and overwhelming competition, online news providers take their search results extremely seriously. As such, the release of Google Panda provided a significant incentive for journalists to improve the quality of their work, reduce their reliance on pre-written content, and add a degree of original analysis or commentary to other people's work. For those of us who had hoped for a solution to internet journalism's "race to the bottom" mentality, Google Panda certainly represented a viable start. As for those in the technology community who wanted to prove that the internet could always find a solution (even to those problems it helped create), Panda also represented a significant win. In fact, the only people who really lost out from Panda were the public relations professionals themselves.

Following the official roll out of the Panda algorithm, Google would later go on to release several new versions of the update including Panda 4.0, which specifically targeted press releases and paid-for "newswire" distribution networks. In the aftermath of this launch, journalists and PR trade publications were quick to proclaim that this was the "death" of the press release, and in some cases, even the death of PR itself. Explaining this reaction on the *Come Recommended* PR blog, Olivia Adams wrote:

> "Google hates content that lacks any sort of merit. This means Google only wants to give higher rankings to

content that deserves it, not because it was paid for. This is why the Panda 4.0 update is going to create more of an issue for PR firms trying to build awareness for their clients." [57]

Given the prevalence of "content that lacks any sort of merit" throughout the field of public relations, this represented a significant blow to the PR community. It appeared that Google – which ironically spends over $9.7 million a year on its own PR – had declared war on the public relations industry. Not only that, but as far as the clickbait headlines were concerned, it had won that war with a single knockout blow. After nearly one hundred years of use (and misuse) the press release was seemingly wiped out overnight. This triumph of technology would provide internet journalism with an opportunity to return to its high quality, independent roots. More importantly, it would also reassure the technoutopians in their belief that the internet will always provide an answer. At least, that's what we all thought.

In actual fact, regardless of what the clickbait headlines claimed, the press release did not "die", but rather returned to its original, more humble role in the reporting process. Prior to the fast-paced internet age, press releases would never have been copied word-for-word by a journalist. In the vast majority of cases, releases were meant to be used as little more than fact sheets around which a story could be built. The press release was rarely meant as a devious device, but rather as a way to provide journalists with a statement that offers the corporate perspective. Rather than relying solely on this statement, it was the job of journalists to collect different versions of events from each of the stakeholders involved (charities, lobbyists, employees, customers etc.), and ultimately establish an honest version of events. It was not until the internet age of real-time reporting and unpaid content generation that journalists came to view the press release as something to be dragged and dropped into a WordPress backend. Press releases did not ruin internet journalism. If anything, internet journalism ruined press releases.

Regardless of the role that press releases may have played in undermining journalism, their supposed "death" has actually achieved very little with regards to improving journalistic integrity.

As we have already seen, the $13 billion public relations industry is extremely versatile, managing a vast armoury of tactics and tools that have grown all the more effective throughout the internet age. Faced with the so-called death of the press release, the PR industry quickly adapted its techniques to take advantage of the new digital landscape.

Following the launch of Panda 4.0, press release distribution site PR Newswire developed a series of tools and guidelines designed to help PR professionals ensure that their content was not being penalised by Google.[58] Likewise, Tom Becktold, a senior marketing president at Business Wire distributions, told industry magazine PR Week that new SEO quality controls had ensured that "our site traffic has not been impacted by the changes".[59] In fact, not only do PR professionals continue to use such distribution services, they have even found ways to turn Google's system of penalisation to their advantage. Through a blackhat SEO process known as Google Bowling, digital PR professionals have found a way to negatively impact the search rankings of their competitors. Ironically, by carefully examining the unethical behaviours that Google will admit to penalising, some PR professionals have been able to create a whole new landscape of unethical behaviour.

In order to understand how such an outcome could be achieved, it is worth taking a moment to examine how modern search engines actually work.

Unlike its predecessors, Google was the first search engine to rely on inbound links as a sign of authority in order to help decide the position of web pages within search results. Inspired by the referencing system used within academic journals, Larry Page developed Google's now iconic PageRank algorithm. By examining the number and quality of external pages that linked to a particular website, PageRank was able to estimate the comparable importance of every page currently listed in Google's database. While PageRank has evolved substantially since its conception in 1996 – now using more than 200 different ranking factors – at the heart of the algorithm still lies a fundamental reliance on inbound links. After growing to understand this relationship between links and search results, digital marketers of the late 90s and early 2000s set about developing numerous search optimisation techniques designed to game this system.

One of the earliest examples of such a technique was the process of "link swapping", in which unrelated businesses would place links on one another's websites in order to appear more popular than they actually were. Moving into the 2000s, this process became so sophisticated that an entire industry was built on so-called "link exchange directories", which allowed users to upload their own corporate URLs in exchange for hosting a few dozen links on an abandoned page of their site. After realising the flaw in their creation, Larry Page and Sergey Brin developed Jagger 2.0, an update for the PageRank algorithm that would ultimately banish the use of reciprocal links. Following this update, if any two sites were found to be linking directly – or even indirectly – to each other, then the two links would automatically cancel out, meaning that neither site would receive the SEO benefits.

Once the process of direct reciprocal linking was effectively destroyed, digital PRs and blackhat SEO merchants turned their focus to increasingly complex forms of link building. One of the most common examples of this trend was the development of so-called "link farms", which consisted of a complex network of websites, all connected via an intricate spider's web of seemingly unrelated links. While Google was easily able to identify directly reciprocal hyperlinks, the aim of these link farms was to add so many layers of abstraction to the process as to confuse the search engine into believing that the links were naturally occurring.

Looking to put a stop to the practice of link farming, Google ultimately began discounting links from websites that contained unnaturally high numbers of hyperlinks per page. Then, as an extra precaution, the 2011 Panda update actually began penalising sites that were found to have purchased unnatural links by dropping them down the rankings. As Google describes the process on its official blog:

> "Any links intended to manipulate PageRank or a site's ranking in Google search results may be considered part of a link scheme and a violation of Google's Webmaster Guidelines... The following are examples of link schemes which can negatively impact a site's ranking in search results:
> - Buying or selling links that pass PageRank. This includes exchanging money for links, or posts that contain links; exchanging goods or services for links; or sending someone a "free" product in exchange for them writing about it and including a link
> - Excessive link exchanges ("Link to me and I'll link to you") or partner pages exclusively for the sake of cross-linking
> - Large-scale article marketing or guest posting campaigns with keyword-rich anchor text links
> - Using automated programs or services to create links to your site" [60]

In some extreme cases, these guidelines have even resulted in brands being removed from Google's search results entirely, often being placed in what has widely become known as the Google Sandbox.[61] One of the most famous examples of such a brand was Interflora, a digital flower delivery service that currently operates in 140 countries around the world. In February 2013, Interflora undertook a massive digital PR campaign, paying bloggers to include links and advertorial content in order to help promote its site. Sensing the unnatural surge in links, Google's algorithm flagged Interflora for unethical practices, ultimately leading to the international brand being dropped from all Google search results. The site was not restored until all suspicious links were removed... eleven days later. Not only did this decision cost Interflora web traffic and potentially tens of thousands of pounds in sales, it also meant that the site was virtually invisible in the build up to Valentine's Day – one of Interflora's busiest times.[62]

At this point, some might wonder how such technologies can be anything but a good thing. After all, by penalising those companies that rely on unethical promotions, surely they are helping to improve the overall quality of content online? Unfortunately, while it may be true that the internet always "finds a way", it turns out that corporate spin is equally resourceful at finding a way around.

Returning to the topic of "Google Bowling", following the above system of penalisation being put in place by Google, several blackhat PR professionals set about turning the system to their advantage. After studying Google's algorithm, they realised that, while an influx of low-quality links could no longer benefit their sites, what it could do was negatively impact one of their competitors. By paying to have a competitor's URL placed on a series of renowned link farms, businesses could (theoretically) force their competitors down Google's rankings – or even have them removed entirely. Considered by some blackhat professionals as just another part of the SEO process, this dirty trick is what has become known as Google Bowling. In fact, the practice is now so widespread that some PR and SEO professionals have even taken to bragging about it within their industry forums and personal blogs. Just a quick internet search for Google Bowling turns up a whole host of articles with informative and unabashed titles such as, "How to steal your competitors' rankings", "Steal your SEO competition's #1 Google ranking" and "Steal traffic from your competitors!"[64]

And it's not just search engines that have failed to weed out the influence of big money and corporate spin. The rise of social media has long been praised for its ability to cut out middlemen, expose corporate lies, and return power to the hands of ordinary people.

As Steve Waddington, Chief Engagement Office at the Ketchum PR firm writes, "anyone with a blog or significant Twitter presence can call an organisation to account... Powerful individuals and journalists have always had the power and the platform to fight back but social media gives everyone a voice". [64]

As we have already seen, this is not strictly true. Most blogs and social media profiles hold virtually no power as they have virtually no following. As a result, they prove effectively useless as a method to "fight back" or even "call an organisation to account". I personally have attempted to use social media to complain to organisations on several occasions, each time receiving little or no real action or response. If anything, my experiences contacting companies via email or phone have always proved far more successful than Facebook or Twitter.

Yet despite this fact, Waddington seems to genuinely believe that social media is somehow liberating people from the shackles of manipulation. As one of the box-out quotes from his book reads,

"The internet has killed spin forever. Transparency is the only possible form of sustainable organisational communication."[48]

Once again, PR spin doctors have been quick to adapt; turning social media technologies to their own advantage. In recent years a new trend in public relations has emerged in the form of online advocacy tactics, or as they are frequently referred to within the industry, "online astroturfing". Astroturfing is the practice of masking a message's true sponsors in order to give the impression that it originated from a grassroots supporter (thus "astroturfing"). In many ways, astroturfing is a process that is almost as old as the public relations industry itself. Prior to the internet, most astroturfing would take the form of front groups, which would act as independent bodies pretending to represent consumers without disclosing their own incentivised agendas. Common examples of such front groups include The American Council on Science and Health – a front group funded by Coca Cola and McDonalds; The Alliance for Better Foods – which is run by the PR firm BSMG Worldwide (whose clients include Monsanto, Procter & Gamble and Philip Morris); The GreenFacts Foundation – which was initially funded by the Belgian pharmaceutical giant Solvay; and the Global Warming Policy Foundation, which works to debunk climate science despite ties to the energy industry.[65]

In many instances the names of these front groups are so Orwellian that they are almost comical. In the same way that the "Ministry of Truth" was used to spin lies in George Orwell's Nineteen Eighty Four, we now live in a world where the Non-Smoker Protection Committee is used to promote the interests of a tobacco firm.[66] Likewise, the Independent Women's Forum is actively opposed to the US Violence Against Women Act.[67] Even the International Food Information Council consists of several former members of both the Sugar Association and the National Soft Drink Association. While they may sound ridiculous, all of these groups actually carry a huge amount of sway, both within the political and the consumer realm. Worst of all is that the age of online transparency may have actually made them more, rather than less effective.

Even as far back as the early 1990s, businesses began to realise that their various front groups could be just as influential online as they were offline – particularly given the perceived lawlessness of the internet at that time. One of the earliest examples of this was from the "independent" International Food Information Council (IFIC). By setting up a website called Kidnetic.com, the IFIC was able to provide educational games, quizzes and fact sheets for the new generation of web-savvy kids. Of course, being a consortium of members from the Sugar Association, the site was careful to include pro-sugar advice such as "cookies, chips, sweetened soft drinks and candy are OK for occasional snacking", and "it's OK to get the huge-size fries or burger when you're really hungry".[68]

While this early example may seem quaint by today's standards, the online astroturfing industry is now more powerful and sophisticated than ever. Corporations are not only creating websites for their front-groups, but are also hijacking the power of social media to promote them. In 2014, the US coal giant Peabody Energy launched its "Advanced Energy for Life" campaign, which was designed to promote coal as a solution to energy poverty in the developing world. According to Peabody, this online campaign prompted "500,000 people to lobby the G20 leaders on the issue of energy poverty"[69] by signing up to the organisation's Facebook and Twitter profiles. What they failed to mention, however, is that the vast majority of these "people" mysteriously appeared on the Advanced Energy for Life social profile in a matter of days. While it would be virtually impossible to prove that Peabody paid for fake followers to boost their campaign, it is worth noting that on one occasion the

Advanced Energy for Life Twitter profile miraculously received almost 85,000 followers in a single night.[70]

While publicly frowned upon by groups such as the PRCA and CIPR, this use of sock puppet accounts is becoming increasingly common within online public relations.

Even within my own experience of the PR industry, it has become quite clear that many businesses still consider manipulation of social media an acceptable practice. In fact, some of the social accounts I used in a previous position are still open to this day. In one instance, I was asked to create multiple profiles on the parenting forum Mumsnet, hoping to use them to build conversations and subtly plug my client's food and drink products for kids. I had a fake name, a fake profile picture, even fake descriptions of my fake kids. While such behaviour would be considered by most as either unethical or just plain creepy, somehow within the online PR industry at that time, it was all just seen as part of the job.

From both my own personal experience and the numerous conversations I have had with other PR professionals, it is abundantly clear that – despite corporate guidelines to the contrary – most PR practitioners have at some point or another found themselves manipulating social media for the benefit of their clients. From fictitious TripAdvisor reviews through to sponsored blog comments, rather than "killing" spin, internet technologies have simply been absorbed into the daily toolkits of corporate interests.

And it's not just businesses that are benefiting from the ease of such spoof advocacy online, various politicians have also been found guilty of attempting to manufacture artificial support through the power of social media. One of the most common examples of such manipulation is the falsification of Wikipedia pages. While the popular online encyclopaedia remains a poster child for the wonders of crowdsourced internet volunteerism, the truth is that Wikipedia is riddled with corporate money and political spin. Advertised as the encyclopaedia that anyone can edit, Wikipedia (along with Wikis in general) has become a breeding ground for what the site's admins politely term "conflict-of-interest editing".[71] In some of the most high-profile cases, these conflicts have included celebrities – such as Naomi Campbell[72] – hiring PR firms to anonymously finesse their profiles, as well as politicians creating multiple sock-puppet accounts in order to trim any potentially negative statements prior to an election.

Once again, these are not isolated incidents, but rather an increasingly acknowledged (if behind closed doors) addition to the digital PR toolset. In fact, the manipulation of Wikipedia has become so widespread that some agencies devote their entire service offering simply to managing their clients' reputations via the site. In 2010, two Texan PR executives set up Wiki-PR, a public relations agency which openly describes itself as "the largest Wikipedia consulting firm".[73] In 2013 the organisation received widespread media coverage after Wikipedia admins discovered that it had set up over 250 sock-puppet accounts to help edit its clients' various pages. Many blackhat marketing organisations use similar "mass account generators" to produce the illusion of support across everything from Facebook groups to LinkedIn discussions.

But it's not just small start-ups that have been caught with their hands in the collaborative cookie jar. In March 2012, lobbyists from Bell Pottinger – one of the UK's largest public relations consultancies – were found to have been editing Wikipedia pages for many of their clients as well as their parent company, Chime Communications.[74] Further emphasising the endemic nature of this issue, when the BBC interviewed Lord Tim Bell, owner of Chime Communications, he claimed that: "you won't find anybody, including journalists, who doesn't do exactly the same thing."[75]

Perhaps the technoutopians at the CIPR and PRCA have a more optimistic opinion of the public relations industry than myself, but with veteran PR executives like Lord Bell making statements like the above, it is hard to believe that the internet provides any strong solutions when it comes to fighting against corporate manipulation and PR spin.

A change in our thinking

For those internet pundits versed in the rhetoric of technoutopia, this chapter will not have listed the internet's failings, but rather its successes. To them, the very fact that such examples exist is proof that the internet's Darwinist approach is working. Bell Pottinger and Wiki-PR were outed thanks to online volunteers. Interflora was penalised as a result of Google's algorithm. Lobbyist front–groups such as the National Soft Drink Association can be easily identified through their listing on the SourceWatch website (a carefully monitored Wiki). But, for every scandal that is uncovered by the internet mob, how many have slipped through the net? Likewise, how many supposed scandals have in fact been nothing more than the ramblings of an angry mob? I am sure Sunil Tripathi and Salah Barhoum would have something to say about the Darwinist nature of the internet lynch mob. Sadly, their stories are rarely told, with such negative experiences struggling to finding space among the optimistic tales of technoutopianism.

As Neil Postman wrote, "it is a mistake to suppose that any technological innovation has a one-sided effect. Every technology is both a burden and a blessing; not either–or, but this-and-that."[76] As with all complex technologies, the internet is a double-edged sword; it can be used to promote truth or to share lies, to spread democracy or to manipulate opinion, to encourage equality or to foster economic injustice. In many ways, such negative uses are not even the result of conscious manipulation, but are simply adverse effects of the complex online ecosystem. When the first unofficial news sites launched in the early 1990s, the independent publishers presumably had no conception that they might one day result in the death of print journalism. Likewise, the early founders of BuzzFeed and Vice could never have predicted how their clickbait format would one day help to spur on a two-by-four effect on consumer attention spans. As much as the free-market billionaires of Silicon Valley may like to position themselves as future-gazing messiahs with some long-term definition of progress, technologists are rarely the oracles they pretend to be. The impact of technology is, as Postman said, "subtle if not downright mysterious",[76] and we must be cautious of those who claim to know its full intentions.

And yet, despite the naivety of these "one-eyed technophiles", we still worship them as the miracle workers of our time. They are the superstars of the twenty-first century, both trusted and loved far more than any democratically elected president or government. These billionaire capitalists have been entrusted with our data, with our communications, even with our relationships. They are the ultimate benevolent dictators. In many respects, we have become so entranced with their miracles that we no longer feel confident in our right to question their methods. We simply accept that they know what is best for us and that – by extension – they should be the ones to define the "correct" direction for the future. As is so often the case with blind faith, this has led to a staggering number of illogical contradictions.

As subjects of a technoutopia, we tell ourselves that billionaires such as Elon Musk and Larry Page genuinely are not interested in money. We look at the world's largest advertising companies

(Google et al) and tell ourselves that they are not interested in manipulating our data for profit. We look at the shock headlines and clickbait articles of once respected newspapers and tell ourselves that the decline of real journalism was simply an inevitability. Rather than questioning the decisions of this elite group, we buy into their promise that all of their decisions – decisions that brought wealth and power into their hands – were simply inevitable parts of humanity's progress – of *our* progress.

As far as these priests of technoutopia are concerned, we live in an age of technological determinism, and any attempt to swim against their tide may as well be an attempt to fight human progress itself. As they see it, even *if* a new technology impoverishes society, costs jobs or causes environmental destruction; if it encourages efficiency and fulfils *someone's* demand, then it will always survive in the free market system. To quote a common piece of deterministic spin used by Silicon Valley entrepreneurs: "you can't ignore the electric lightbulb just because it puts the candlemakers out of work". To them, their products *are* progress, and progress is always inevitable. Driverless cars *will* put taxi drivers out of work. Robotic assembly lines *will* take factory workers' jobs. Aerial drones *will* put delivery companies out of business.

To a true technoutopian, the issues described throughout this chapter *will* happen, and anyone who attempts to fight such progress – be they governments, legislators, unions or citizens – is simply fooling themselves. If blogging and cheap media were destined to destroy traditional journalism, then that is what will happen. If the internet was destined to redistribute wealth into the hands of fewer and fewer corporations, then who are we to question it? If PR and advertising is to expand and increasingly intrude on our personal privacy, then that is simply what market forces require.

This is the view of the Silicon Valley elite, but it is not a view that

I believe we are at liberty to leave unquestioned. By hiding behind such bold and deterministic rhetoric, these entrepreneurs are able to pretend that they hold some sort of monopoly on the future of humankind. They stage-manage progress on a day-by-day basis while simultaneously telling the world that the direction of progress cannot be changed. But the future is not set in stone, and we too can define the direction of progress just as they have. As a society, we must learn to understand that just because something is new does not mean it automatically represents progress. Likewise, just because something is more efficient does not automatically make it better than what came before. Progress is defined by the choices of citizens and governments just as much as it is by entrepreneurs.

If Google had decided to avoid advertising (as originally planned) and provide its search engine as a paid subscription, we may have all ended up using a very different kind of internet. Likewise, if lawmakers had truly believed the non-interventionist rhetoric of so many Silicon Valley entrepreneurs, we would all be wandering around with 3D printed firearms right now. After all, if entrepreneurs like Mark Zuckerberg and Sean Parker can honestly believe that job cuts and music piracy fall within the bounds of inevitable "progress", why can't the mass production of unlicensed firearms? The reason, is that governments and lawmakers are sensible enough to realise that all new technology comes with a potential societal impact. As a result – no matter how much it pains the neoliberals of Silicon Valley – all "progress" has to be managed and regulated to ensure that its benefits (efficiency, profitability, consumer value) are not outweighed by its potentially negative impact upon the planet or society.

Nothing about the internet is inevitable. Every new website, new application, even new law, helps to shape the direction of the overall system. The internet does not inevitably make the world more equal, transparent or democratic. None of these things come naturally to any system, no matter how vast or theoretically open. They are instead complex social ideas which can only be fought for.

The internet provides everyone with a voice, but it does not provide everyone with equal representation. It provides citizens with the ability to share news, but it does not teach them how to be journalists. It provides openness through information abundance but loses transparency through information glut. It encourages equality of access but widens the inequality of wealth. It removes the gatekeepers of journalism, only to replace them with the gatekeepers of PR. To quote the title of Andrew Keen's 2015 book, the internet is not the answer.

The myths outlined in this chapter are truly pervasive throughout our technoutopia. Visions of the internet as intrinsically tied to equality and democracy are shared by technologists, politicians, activists and even the public as a whole. But while many of these myths may have arisen naturally, they are also increasingly spurred on by those that benefit most from their widespread acceptance.

As much as the internet was destined to remove elite intermediaries from the public sphere, there is now a growing layer of gatekeepers emerging throughout the web. Unlike their offline predecessors, these gatekeepers are rarely called to account as their influence is simply too engrained within the functions of the net itself. They define what we see, read and hear. They own our secrets, our data and increasingly our social lives. They are the search giants. The new kings of technoutopia.

The internet removes bureaucrats, intermediaries and gatekeepers

> "The real-time nature of the internet amplifies discussion and democratises communication so that everyone has an equal voice, meaning that a more stringent editorial function is arguably more important than ever."
>
> Steve Waddington

> "Disintermediation is as much mythology as fact. Its effect is to make the new mediators - the new gatekeepers - invisible."
>
> Eli Pariser

How many technologies do you use every day? For most of us, the modern world is a staggering array of technological advancements, all crying out for our attention and bombarding our brains with information and images from an endless selection of flickering displays.

When waking up in the morning, the very first thing I do is pick up my mobile phone and switch off my alarm. Then, before I even realise what has happened, I find that ten minutes have passed and I've been sat scrolling through Facebook or Twitter. I often think that from a psychological perspective, free alarm clock apps are one of the cleverest things to have been bundled with smartphones. While perhaps not the most interesting of applications from a technological perspective, the inclusion of an alarm clock sets the precedent for the rest of our mobile-obsessed day. We wake up, and then we *have* to pick up our phones. We go to bed, and we *have* to pick up our phones. Our daily routine has been top and tailed with screens.

Having picked up my mobile phone in order to switch off the alarm, I quickly find myself spending some time mindlessly looking at it. I open my web browser to check my gmail account, maybe watch a few videos on YouTube that my friends have shared with me. Then I might check my calendar for the day, perhaps even casually browse the analytics of my Blogger account for a bit of an early morning ego-massage. Sometimes I will check my journey on Google maps or conduct a quick internet search for the day's weather report.

In some ways it's quite scary to think that I access around 15 different technologies before I've even got out of bed in the morning. Thanks to a seemingly innocuous decision – to include a free mobile alarm clock – my eyes open, and then they hit a screen. What's even scarier is the fact that all of the technologies mentioned above, have come from a single corporation – Alphabet Group; or as it's better known, Google.

Here's a little summary of the above routine again, this time, with brand names in place: I wake up and quickly pick up my Google Pixel phone. Navigating the Google Android operating system I switch off my native Google-developed alarm clock app. Moving on to the home screen, I open my Google Chrome browser to access the internet, which launches my Google Search homepage in the process. After navigating to my Gmail account, I watch a few videos my friends have shared from YouTube – a once independent video site subsequently snatched up by Google. Next, I visit my Google Blogger account, checking the Google Analytics add-on for an early morning ego boost. Finally, I check my journey for the day on Google Maps before having a quick scour of the Google Weather add-on, which I installed to my Google Chrome web browser.

Google owns my life. It owns my data, it owns my routine, and quite frankly, I'm not sure I would survive that long without it. As the company's wealth, and ultimately buying portfolio expands, Google will also control increasing swathes of my life outside of the computer screen. Soon, many predict that Google will become one of the world's most powerful internet service providers,[1] currently wowing the world with its ultra-fast Google Fiber and Google Loon projects. At the same time, Google is increasingly invested in entirely unrelated sectors, such as automotive production (leading the race for of the world's first commercial driverless cars), as well as military applications such as the purchase of Boston Robotics – a company specialising in the production of the first armoured war robots. But it's not just companies owned by Google that are increasingly taking hold of our lives, there are also the 300+ businesses that Google holds some form of significant stake in. Through the various investments made by the Google Ventures arm of its business, Google now holds influence in transportation (Uber), 3D printing (Caron3D), home automation (Nest), instant messaging (Slack), food and beverages (Blue Bottle Coffee) and even healthcare (Foundation Medicine). Through this eclectic and ambitious investment portfolio, Google is set to become an even more intrinsic and inescapable part of our daily lives.

But is that really such a bad thing? While the arguments presented in the previous chapter may have shown that internet technologies have done little to improve transparency or genuine equality, it seems safe to assume that the products presented by tech giants such as Google have revolutionised our lives. The mere fact that I use ten or more Google products before getting out of bed in the morning is a testament to their invaluable impact. As a result, how could more of Google's miraculous inventions be anything but a good thing?

While the wondrous, imaginative, and often downright mysterious goings on at Google HQ may bamboozle us with endless joys, they come with a very stark contradiction to the wider mantra of internet freedom. One of the most enduring virtues of internet technology is its ability to remove gatekeepers and place power and responsibility back in the hands of ordinary people.

In the internet age, gatekeepers are a sign of needless bureaucracy and of money grabbing middlemen out to make a fast buck off of the hard work of genuinely creative people. In opposition to this, eBooks and Amazon have successfully cut publishers from the publishing industry; Soundcloud and Spotify have cut agents and record labels from the music industry; YouTube and smartphone cameras have cut video production companies from the filming process; eLearning courses have cut teachers from the teaching process; while bloggers and "content marketers" increasingly banish wealthy owners, editors and investigative journalists from the process of professional journalism.

While internet technologies may have been successful in cutting out the middlemen from so many professional fields, an entirely new class of gatekeepers is beginning to emerge in their place. These gatekeepers are often less professional, less accountable, and yet in many ways more powerful than their predecessors could ever have hoped to be. And yet, somehow, throughout this process, the loyal proponents of technoutopianism remain willfully blind to the contradiction. They spout the buzzwords of disintermediation and "digital disruption", calling for the end of traditional gatekeepers while simultaneously worshiping (or even becoming) gatekeepers themselves.

When it comes to such delusions, Google certainly knows how to talk the talk. Its marketing message is one of quirky individuality, expressive liberty and freedom from the stuffy boardrooms of bureaucratic corporate management. And yet, as the company's spin doctors and marketing spokespeople showcase how perfect the world would be if we all embraced the marvels of disintermediation, Google is quickly becoming the greatest gatekeeper of them all.

Gatekeepers: An end to the elite

Before looking at how companies such as Google became such powerful gatekeepers, it's worth taking a moment to examine what it is that makes internet evangelists so naturally hostile towards the prospect of traditional intermediaries.

The crudely named process of "disintermediation" has been intrinsically tied to internet technologies ever since the early days of the World Wide Web. Even the very first uses of email and Usenet forums were seen as a way to circumvent traditional media and communication outlets, effectively allowing first-generation nerds to cut out the corporate middleman. As the internet moved into the two-way communication of Web 2.0, those savviest of organisations began to realise that disintermediation represented both a potential opportunity and a significant threat.

For new-age internet gurus, it highlighted a great marketing opportunity, allowing them to position their web-based businesses as the creative solution to decades of bureaucratic communication and corporate money making. As Ted Leonisis, then Services President of America Online, commented, "Why should the media be allowed to filter your message anyway? In the near future, everybody will have access to all the information they need to make their own decisions."[2] Combined with the psychedelic cyber-speak that surrounded computers at the time, this libertarian, anti-establishment rhetoric played perfectly to the ambitious, pill-popping pioneers of the early nineties internet. From this moment onwards, the internet would forever be associated with freedom and individual choice, with those who attempted to control access (for better or for worse) being branded as power hungry gatekeepers attempting to subvert information's natural desire to be free.

As the internet evolved, the definition of these gatekeepers expanded to include anything or anyone that undermined the principals of equality that had come to define the era of blogging and Web 2.0. In order for such technologies to function – at the level that technoutopians had optimistically promised – the public needed to be convinced that everyone in the world was capable of anything they wanted, as long as the technology was there to support them. In order to be a journalist, you did not need years of training in news production, geopolitics or the ethics of journalism; just a smartphone camera and a free YouTube account. Likewise, to be a taxi driver, you did not have to learn the street names around London, or understand how to calculate fares in a fair but profitable way; you just needed to install the Uber app and follow a built-in sat nav.

All internet technologies are designed to convince us that we, as individuals, are capable of anything we want. Want to set up a shop? Start an eBay account. Want to learn a language? Install an app. Want to fix your own car? Watch this video tutorial. While this may sound empowering, in reality it does not convince people that they can *achieve* anything, but rather that they can *do* anything – with or without any real commitment to the end cause.

This mindset fitted perfectly with the notion of equality that had come to dominate the Web 2.0 internet. The only problem was, that as long as there were real life experts in the world, the whole system was undermined. As a result, the technoutopian community – either knowingly or as a natural libertarian process – set about to attack the very notion of the "professional class". To be a professional was to hold a skill that others could not easily possess, and to even admit that such a skill existed within the internet age was to be branded an elitist.

This was the start of the war on professionalism. When internet entrepreneurs such as Andrew Keen attempted to argue that the "cult of the amateur" could never replace professional journalism, he was immediately dismissed as an elitist.[3] Likewise, when lecturers attempt to argue that massive open online courses (MOOCs) do not provide students with the stimulating environment necessary to truly engage with a subject, they are fobbed off as fusty academics, desperately attempting to justify their cushy jobs and their inflated salaries.[4]

But even if traditional elites are out to protect their own jobs, that does not undermine the fact that professionalism is a vital part of our social and economic society. As much as internet evangelists are loathed to admit it, to become a professional takes more than just access to certain tools. In the same way that providing paints and a canvas does not make someone an artist, providing a platform to publish news does not necessarily make them a journalist. To be a professional takes inspiration, experience and commitment. Unfortunately such logic does not fit with the narrative of universal equalisation that remains so prevalent online.

Strangely enough, however, when it comes to the private sector – in particular, those that have embedded themselves in the rhetoric of technoutopia – the internet community has an oddly contradictory relationship with the notion of elitism. Individuals such as Larry Page and Sergey Brin, who sit in their Silicon Valley mansions, being whisked around in driverless cars and owning, not just private jets, but also their own private airports,[5] are somehow exempt from the anti-elitist rhetoric that surrounds their offline competitors. As Andrew Keen writes in *The Internet is Not the Answer*, "rather than more openness and destruction of hierarchies, an unregulated network society is …compounding economic and cultural inequality, and creating a digital generation of Masters of the Universe".[4]

These "Masters of the Universe" have positioned themselves as the gods of technoutopia, allowing them the privilege to not only draft the rules of play, but also to ignore those rules as and when they see fit. In this way, the leading entrepreneurs of Silicon Valley are able to publicly declare that the internet encourages equality, while simultaneously raking in personal fortunes that would dwarf any oligarch of the pre-internet century. They tell us that the internet represents individual control, while openly selling our privacy to the highest bidder. They disdain government interventionism; while simultaneously benefiting from tax-educated employees and sponsored research grants. They spout the democratising effects of their products, while openly spending tens of million on government lobbying. These are the contradictions that define a technoutopia; a blind faith that technology – and technologists – always want what is best for humanity.

And yet, like the followers of a spiritual cult, the citizens of technoutopia have too much faith in their leaders to even notice that such contradictions exist. Those that do are branded pessimists and Luddites and are accused of attempting to hold back the inevitable march of progress. What we do not realise, however, is that these are the people who define the very vision of progress towards which we are marching. As the media theorist Douglas Rushkoff writes, "our interpretation of what it means to progress might have been formulated for us by those at the top of the pyramid, who stand to profit from our mindless participation."[6]

And so, we gleefully follow along with this "mindless participation". We damn elites in exchange for some unattainable vision of perfect equality; even if it comes at the expense of proficiency and professional vigour. And all the while, an elite group of uncontested billionaires stand upon their silicon mountain, somehow able to remain outside the realms of public scrutiny. With a smile on their faces and a personal wealth of $39.4 billion (as in the case of Sergey Brin), they tell us – the merry citizens of technoutopia – that we no longer need to live in a world that tolerates elitism.

An attack on inefficiency

While the strive for some idealist vision of equality has certainly proved a keystone in the internet's attack on traditional gatekeepers, it is not the only factor to have driven this trend. By acting as middlemen between what the public wants and what the public is able to achieve, traditional gatekeepers undermine two of the founding principles of a technoutopian society. Firstly, they imply that the public cannot do everything themselves – regardless of the number of apps, YouTube tutorials and internet tools provided. This, in turn, questions the universally accepted notion that the internet is an unparalleled equaliser. Secondly, (and perhaps even more offensive to utopian sensibilities) they act as middlemen, supposedly adding an extra layer of inefficiency to an otherwise streamlined world.

Within a technoutopian society, inefficiency is the original sin. Having come to worship the ill-defined notion of progress, a robot-like devotion to efficiency is widely seen as the highest human value. Governments must crack down on inefficient spending, businesses must cut the "deadwood" from their workforces, even individuals must search for ways to increase the efficiency of their daily leisure time. Satellite navigation systems provide us with the most efficient ways to get to work; predictive texting helps us to decide what we want to say before we've even said it; algorithm-based dating sites help us to find love without "all the stress" of traditional dating. In an information rich yet time poor attention economy, efficiency has become the ultimate goal of our society. If something is not running at 100% efficiency, then it simply isn't finished yet.

The possibility that inefficiency is synonymous with variety, and as a result could be seen as a positive value has subsequently become unthinkable. In reality, however, individual enjoyment rarely comes from efficiency, but rather from the challenges that human beings face as a result of *inefficiency*. As Nicolas Carr writes in his 2015 book *The Glass Cage*, "We're happiest when we're absorbed in a difficult task, a task that has clear goals and that challenges us not only to exercise our talents but to stretch them."[7] As much as we may receive pleasure from the end result of such a task, it is the task itself that makes us happy. On a personal level, writing this book has proved extremely challenging, and at many times I have looked forward to the end result and wished for a simpler way to get there. If, however, a machine had existed to write the book on my behalf, I do not imagine that I would be nearly as pleased with the end result. It was the challenge of writing it that made the whole process so enjoyable and the end result so satisfying. As our obsession with efficiency grows, the opportunities for such enjoyment will become ever fewer and farther between.

While the process of writing a book may not yet have fallen foul to the efficiencies of automation, we are not very far from a future in which such an idea could become reality. In a similar vein to the trashy novel-writing machines from George Orwell's 1984, software algorithms created by Professor Philip Park of the Insead Business School have so far successfully written over 200,000 instructional books, which are currently available to buy on the Amazon book store.[8] While the algorithms behind these robot authors remain in their early stages, the creators of this project do not rule out the possibility that their algorithms could one day successfully draft a work of fiction. Similar efforts are also being pursued in the field of journalism, with the Guardian's Open 001 service being launched as the world's first robot collated print newspaper.[9] At the opposite end of the spectrum, services such as Blinkist look to maximise the efficiency of our reading, skimming books on our behalf and then providing a short digestible summary for our enjoyment.[10] This is the bizarre extent to which efficiency has infiltrated our world. To save time we delegate the writing of our books to algorithms and then ask robots to read them on our behalf.

Of course, it cannot be denied that the introduction of such automated processes has helped to improve efficiency, but in doing so they strip the joy from human endeavours. It's only by redefining writing – or even reading – as a chore, that improved efficiency becomes a necessary goal. In this way, so many of life's little pleasures have been stolen and perverted by the endless search for ultimate efficiency.

Why should anyone explore their home town when a satellite navigation system could just as easily get them from A to B? Why spend time learning a new language when Google Translate can interpret for you in an instant? Why pick out flowers for a loved one when sites such as MoonPig already know the "best" combinations based on previous browsing histories? These are the simple pleasures of *doing* that are increasingly lost to the technological age.

Those that argue such a case, however, will likely be branded as little more than sentimentalists; radical hippies who are simply unwilling to accept that to fight new technology is to fight the future itself. But in taking such a deterministic stance are we really doing what's best for society? Equally important, are technology firms even doing what is best for their customers? Increasingly, the entrepreneurs of Silicon Valley are sounding less like the revolutionary libertarians of the internet age, and more like the economic corporatists of the late twentieth century.

In its obsession with efficiency and progress, the rhetoric of Silicon Valley has become less focused on individual freedom and choice and increasingly obsessed with market freedom and deregulation. In a world where the future is apparently decided in advance, and any deviation is simply "fighting the inevitable", what room is left for the idea of individual choice? Can the organisations and entrepreneurs that support such a deterministic notion believe that freedom of choice has any role to play in the future of our world? By accepting the idea that "progress" is inevitable, the entrepreneurs of Silicon Valley are essentially rendering the concept of legislation, government intervention and even consumer protection as irrelevant. If such a worldview is accurate, what would be the point of attempting to legislate new technologies? Particularly if their rise to popularity is already set in stone.

But perhaps that's the point. A cynic could argue that the business leaders of Silicon Valley do not really believe in such free-market idealism and are simply using this lofty rhetoric in an effort to open up their markets and avoid the profit-squeezing restraints of government regulation. Personally, I believe that the motives of these utopians go far beyond short-term profiteering. Anti-government rhetoric has grown so endemic throughout the technology community that it is hard to believe that such sentiments are nothing more than an attempt to pay fewer taxes. When Mark Zuckerberg says that Facebook could play an instrumental role in solving the crisis in the Middle East, all of the evidence suggests that he honestly believes it. He believes that, somehow, digital technologies will always find an answer, even if it is an answer that we cannot yet perceive. Likewise, when technology journalists such as Andrew Sullivan proclaimed Twitter to be "the critical tool for organising the resistance in Iran" (despite only 0.027% of the population actually using

Twitter[11]), I think that he honestly believed it too. The use of such lofty rhetoric has not become normalised as a result of marketing spin (although I am sure that plays some part), but rather as the result of a genuine passion and overwhelming optimism that flows throughout the industry. That is the most fascinating – and possibly scariest – thing. These people genuinely *believe* in what they are selling.

Several times throughout this book I have described technoutopianism as closer to a religion than an industry movement, and at this juncture it should be easy to see why. Total faith in technology has not only provided those of Silicon Valley with something to devote their lives to, it has also provided a near biblical level of optimism and hope. Presumably, for those that truly believe, knowing that sooner or later some undiscovered technology is going to solve all of the world's problems must provide a similar level of comfort to the notion that some benevolent force or unseen paradise is waiting for us in the life to come.

Sadly, while this technological devotion may bring its tenants tremendous joy, for the rest of society this determinism has manifested in unstable free market economics, destruction of jobs and the increasing mechanisation of personal leisure time.

In this way, despite their laid-back cultures and public facades, Silicon Valley gurus have far more in common with free-market corporatism than they do with liberal hippie countercultures. From the removal of bureaucratic and elitist gatekeepers, to the wider obsession with technological efficiency, so many of Silicon Valley's beliefs are little more than carefully rebranded relics of the capitalist age.

In the same way that the mass production techniques of Frederick Taylor or Henry Ford turned interesting and engaging work into a series of repetitive actions, digital technologies have turned all of life's activities into little more than a McDonalds' production line.

All of the enjoyment, the variety, and yes, even the challenges of modern life have been stripped away in the name of efficiency. The biggest irony, is that while Fordism offered the tangible social benefit of mass job creation, Silicon Valley's efforts have succeeded in inspiring the opposite effect. Digital technologies have taken efficiency to a level in which human beings are essentially surplus to requirements – even in those tasks that were scarcely considered a chore to begin with. As work becomes ever less complicated and our lives grow increasingly automated, not only does the need for skilled or passionate workers reduce, but we begin to call into question the need for a physical workforce altogether. One only needs to survey the employment records of Silicon Valley's biggest and best to see that such a prediction is not some dystopian vision for the future, but rather an observable effect of our current technological landscape.

Since the rise of Amazon, the UK has lost nearly two-thirds of its high-street book stores. Likewise, following the popular growth of iTunes and Spotify, music industry revenues have more than halved. And it's not just in the United Kingdom, US figures also highlight a 45% reduction in the number of professional musicians, with internet giants centralising music distribution and closing traditional avenues to success.[4] And that's just the beginning, according to one University of Oxford report, the UK is expected to lose up to ten million jobs as a result of technological "progress" in the next twenty years.[12]

The real question, however, is whether or not those jobs are being replaced. For those that are not utopian enough to push the argument of inevitable progress, defenders of Silicon Valley often fall back on the idea that internet technologies such as Amazon and Google are actually creating more jobs than their products destroy. Unfortunately, this simply isn't true.

As Andrew Keen highlights in *The Internet Is Not The Answer*, while Google is currently estimated at around seven times the size of General Motors, the search giant still only employs less than a quarter of GMs workforce.[4] Likewise, despite being acquired by Facebook for $1 billion, the photo-sharing site Instagram had only thirteen official employees at the time of its acquisition.[13] Instant messaging platform Whatsapp offers a similar example. Having been acquired by Facebook for $19 billion in 2014, the company only actually boasted a total of fifty-five employees – hardly a booming job creator.[14]

These companies all make innovative products and more often than not provide great services to their customers. Sadly, when it comes to the creation of wealth, their business models naturally compound ever higher volumes of wealth into the hands of an ever smaller number of people. And all the while, more traditional businesses such as Kodak and British Telecom are being forced to slash hundreds of thousands of jobs as their customers switch to their "free" online competitors. Worse still, these competitors are not really free at all; they are simply more willing – and able within the lax online environment – to sell customers' data, in a way that traditional companies such as British Telecom could never dream of getting away with.

At this point, those versed in the arguments of technoutopia will likely fall back into the default position of technological determinism. To return once more to that carefully crafted piece of Silicon Valley spin, "there's no point banning the electric lightbulb just because it could put the candlemakers out of work." In this simplistic regard, the utopians have got it right. In a competitive market, any technology that improves upon its rivals will almost inevitably succeed (with a few exceptions). Nobody in their right mind would consider banning technologies that help to make the world a better, safer or more environmentally-friendly place. But this is not to say that we should not manage and assess the impact of those technologies before they are fully integrated into our lives. If anything, that is our duty as a part of a caring and democratic society. As Michael Harris puts it in his book *The End of Absence*:

> "If we work hard enough to understand this massive game changer [internet technology], and then name the parts of the new game we want to go along with and the parts we don't, can we then pack along critical aspect of our earlier lives that those technologies would otherwise strip from us?".[15]

This is how we need to approach all technologies that could have potentially unseen cultural or social effects. Unfortunately, such an undertaking involves a significant amount of thought, patience and public intervention, three things that are in short supply in the internet age. To our free market Silicon Valley entrepreneurs, such attributes are considered synonymous with bureaucracy, inefficiency and the restraint of market freedoms. And as we already know, within a technoutopia, devotion to efficiency and the removal of bureaucratic gatekeepers are the highest possible goals; goals that would trump social or societal wellbeing any day of the week.

But in our drive for efficiency, and march towards inevitable "progress" (whatever that means) are we really making the world a better place for human beings to live in? More importantly, is this cut-throat free market approach even achieving the goals that the utopians themselves set out to undertake? Have we really abolished the gatekeepers of old media, or simply replaced them with a group of more powerful, yet less accountable, "Masters of the Universe"?

Possibly the biggest problem of truly free markets is their natural predisposition towards unregulated monopolies. As the former

Times journalist Simon Jenkins writes "nobody but a fool believes that a free market in anything, left to its own devices, will tend to perfect competition. Economic history attests that it tends to monopoly. That is why it must be regulated. Such regulation, in every sphere of economic life, is democracy's most onerous but essential responsibility."[16]

The internet has proved the ultimate playground for free market economists, looking to test their theories of disintermediation and deregulation in an environment which, until very recently, was generally considered outside the law. As their dream has turned to reality, and the internet has grown to become one of the dominant forces in all of our lives, free market competition has quickly reverted to demagogy.

With largely unregulated organisations growing so vast, and with such unprecedented opportunities for economies of scale, these new gatekeepers of the internet age have come to bend the market to their will, ultimately closing out competitors and the potential for true freedom of choice.

As David Shenk insightfully predicted back in 1994, "in the increasing pace and distraction of our informationized culture, our two centuries of democracy will fall prey to demagogues"[2]. Or, to put it in the plainer terms of Upworthy founder Eli Pariser, "disintermediation is as much mythology as fact. Its effect is to make the new mediators – the new gatekeepers – invisible."[17]

So who are these invisible gatekeepers that have come to control the internet age? Well, for the most part, they are the same as those of the traditional media landscape. They are an unstructured group of elites, entrepreneurs and corporations that have taken the free market values of technoutopia to their logical – if extreme – conclusion. These innovators and "digital disruptors" continue to filter information and individual access just as their forefathers did in the era of print journalism. Now, through careful monitoring and algorithmic recommendations, they decide what is best for us to see. Behind the scenes they regularly destroy jobs in the name of individual profiteering, and will happily limit genuine choice in exchange for tightly controlled "variety". But you will not find these individuals lurking in dimly lit boardrooms or the shadowy corridors of power. More often than not, they can in fact be found squatting on your browser's favourites bar, or even installed as an application on your mobile phone. They are the Amazons, iTunes, Googles and Ubers. They are the new rulers of the digital economy, and, to use Andrew Keen's phrase, they are typically nothing more than those of the old industrial economy – "on steroids."

Amazon – A new generation of gatekeeper

For the average internet user, services such as Amazon or iTunes are not gatekeepers, but are simply helpful applications designed to provide quality goods and services at a price lower than that of the traditional high street – and rightly so. I have no qualms with the quality of services these companies provide, nor with their rightful success. Despite some of the anti-corporatists rhetoric of previous chapters, critics would be wrong to assume that I disagree with the growth of corporatism, or even capitalism in general. There is no Marxist agenda here; only a wish to understand how the internet has impacted us as both individuals and as a society in a frank and honest way.

Through utopian rhetoric we have become convinced that the internet is unquestionably an empowering and democratising force. Technology evangelists point to its role in cutting out middlemen and weakening the institutions of power. They claim that we are working to replace elitists and corruptible politicians with a decentralised democracy, an open space where "everyone has a voice".

As the standard narrative goes, thanks to internet technologies anyone with even limited technical experience can build a website, start a blog, or even launch their own business online. That was the dream anyway. A dream that, for most, collapsed as early on as the 1990s.

Internet technologies may have resulted in genuine disintermediation, but for every intermediary that has been removed, the remaining gatekeepers have grown in both size and power. What were once the values of disintermediation – functionality, convenience, familiarity – have grown synonymous with the pitfalls of a corporate monopoly. Now, with every middleman that the internet cuts out, individual choice and empowerment (the very notion of a buyer's market) are also slashed.

Instead of decentralisation, the Wild West of internet economics have afforded a small selection of gatekeepers full control. As Eli Pariser puts it in his illuminating critique of internet technologies, *The Filter Bubble*, "people who are renting and leasing apartments don't "go direct" – they use the intermediary of Craigslist. Readers use Amazon.com. Searchers use Google. Friends use Facebook. These platforms hold an immense amount of power."[17]

Possibly the best example of this is Amazon, the multi-billion-pound ecommerce and cloud computing giant. Originally founded by Jeff Bezos as an online bookstore in the early 1990s, by 2015 Amazon.com Inc had surpassed Walmart as the largest and most recognisable retail brand in the US. Bezos himself is a classic example of a technoutopian, comparing the development of his empire to the work of a "missionary",[18] while simultaneously using Amazon's enormous market power to crush competitors, pressurise suppliers and lobby governments for reduced taxation – further cementing the company's monopolistic power.

According to the libertarian view shared by much of the internet community, "digital disruptors" such as Amazon should not be held back by stuffy governments and bureaucratic regulations. Instead, they should be left to innovate and improve within a free market structure. So how has that been working out for everyone's favourite online retailer?

So far, not so well.

At this point in time, Amazon has been embroiled in so many controversies around the world that it has even been given its own 7000 word "Amazon.com controversies" Wikipedia page.[19] Adopting the approach of corporate giants such as Walmart and Starbucks, Amazon has undertaken a land grab of market share by using its size to drop prices below competitors (at great cost to Amazon itself), and then simply playing the waiting game until offline retailers are forced out of business. As Andrew Keen describes it, "the bigger Amazon has become, the cheaper its prices and the more reliable its services, the more invulnerable it has become to competition."[4] Or, as CEO Jeff Bezos himself put it, Amazon's negotiators go after smaller businesses "the way a cheetah would pursue a sickly gazelle".[20]

But it's not just Amazon's competitors that must suffer under the company's free-market race to the bottom. Even those businesses that supply and stock Amazon's virtual shelves are considered fair game within the cut-throat online economy. To list just a few examples, in recent years Amazon has come under criticism for telling publishers and authors that they were not allowed to sell discounted books via their own websites. This was most clearly highlighted in the case of the French book publisher Hachette, which launched a high-profile dispute against the ecommerce giant for the right to set its own prices and discounts.[21]

In May 2014, Hachette refused to give Amazon pricing control over its ebooks, which would have seen most of the publisher's digital titles discounted to less than $10 per book. In response to this, Amazon used its significant power and influence to undertake "aggressive negotiations" - a series of predatory business tactics that resembled those of a 1920s mob shakedown. These included preventing customers from being able to pre-order Hachette titles on Amazon, removing discounts offered on Hachette books and even going so far as to stall shipments of the publisher's titles for weeks at a time. These tactics succeeded in placing enormous pressure on Hachette, ultimately slashing their sales throughout the negotiation process. By the end, over one thousand authors signed a petition demanding that Amazon stopped "using writers as hostages" in its negotiations.[22]

Third party suppliers have also felt the full force of Amazon's predatory tactics. Postal services such as United Parcel Services and FedEx are being squeezed by the ecommerce giant, with Amazon using its economies of scale to beat down regular delivery prices – something that smaller retailers could never hope to achieve.

As journalist and political activist Jim Hightower writes, "[Bezos] goes for the throats of both large and small businesses that supply the millions of products his online behemoth sells. They're lured into Amazon by its unparalleled database of some 200 million customers, but once in, they face unrelenting pressure to lower what they charge Amazon for their products, compelled by the company to give it much better deals than other retailers can extract."[23]

But is this really all that problematic? The way Jeff Bezos sees it, Amazon is simply offering a platform "where consumers can find and discover anything they want to buy online".[24] Better still, Amazon is doing it at a price and an efficiency (there's that word again) which outperforms anything else on the market today. Near endless choice, offered in a convenient way at a price that the average consumer could actually afford. Can that really be such a bad thing?

Amazon does offer great value, and a generally great experience, but does it offer consumer choice, or is it simply the illusion of choice? The seemingly endless virtual shelves of Amazon.com offer greater *variety* than anything that has come before, and yet, through its monopolistic business practices, it is simultaneously limiting genuine *choice*. That is possibly the most ironic thing about internet technologies: the very thing they promise – freedom of choice – is the very thing that services such as Amazon, Uber and Google seem set on destroying.

By forcing so many of its competitors (and even suppliers) out of business, Amazon is ultimately limiting the range of genuine choices available for its customers. Worse still, through its enormous market share, Amazon is now able to openly limit choice, wiping out entire brands and product lines with the press of a button. Possibly the most controversial instance of this is Amazon's refusal to sell products that directly compete with its own. As an example, try searching on Amazon for the Google Chromecast, a popular plug-and-play digital media player. Is the top result for this simple and straightforward search the product you were looking for? No. The top result is, of course, the Fire TV stick, Amazon's own digital media player and direct competitor to the Chromecast. Not only is Google's rival product not listed as the top search result, it is not listed at all on the site. Amazon has effectively wiped the device from its database. The same is also true of Apple TV, another of Amazon's leading competitors.

And who is to say that this purging of competitor products will stop at digital media players? Amazon already has a noticeable lack of e-readers online, who's to say that this is not an active attempt to promote the Amazon Kindle? With the retail giant now expanding into smartphones, tablets and other portable devices, how do we know that they will not suddenly cut iPhones, Blackberry's or alternative Android phones from the store? Already, a search for "tablet" on Amazon lists the Kindle Fire (the retailer's own tablet) above significantly more popular Apple and Android alternatives.

But is this fair game? As far as Jeff Bezos is concerned, organising the company's products in this way is Amazon's prerogative. In the same way that nobody can force a local shopkeeper to stock specific goods, nobody can tell Amazon that it has to stock certain competitor products. Likewise, the decision to place Amazon goods at the top of search results is no different from a high-street retailer deciding to give their in-house products greater shelf space.

The only problem with this quaint analogy, however, is that Amazon is not some local grocery store, it is a $300 billion company which has become the go-to portal for almost everything we, as a society, consume. With that position of power must come a certain degree of responsibility.

According to figures from 2013, in the US alone, Amazon's online sales totalled more than its closest nine rivals combined – including the likes of retail giants such as Apple and Walmart. With this in mind, the idea that the internet has removed gatekeepers comes with a deep sense of irony. Amazon now has more monopolistic power than any offline Starbucks or Walmart could have ever dreamed of. Even the worst offenders could not get away with openly removing access to their competitors' products, particularly off the back of an excuse as flimsy as preventing "consumer confusion"[25] (yes, that is the actual reason given by Amazon).

And it's not just competitor products that are being censored by the online retail giant. Amazon has also made a nasty habit of removing content from its Amazon Kindle e-reader without prior notice or explanation for customers. For anyone who has read the works of George Orwell, the move came as particularly creepy, with accusations of Amazon attempting to rewrite history or ban books plastered across the papers. In one example of such censorship, Amazon automatically deleted all Kindle books uploaded by Australian publisher MobileReference, which just so happened to include George Orwell's 1984 and Animal Farm. The irony of this move was not lost on censorship activists.[26]

While these cases were typically nothing more than poorly-handled responses to copyright infringement or attempts to remove genuinely offensive material, they still highlight a worrying degree of power wielded by this monolithic information gatekeeper. For a world where e-books could one day come to replace printed texts, the sheer fact that companies such as Amazon have the ability to wipe books from existence should be a point of ethical discussion and potential political concern.

Of course, in the true spirit of a technoutopia, libertarian internet evangelists would argue that it is not Amazon's job to consider such things. As with Google, Amazon does not really "do" politics. They are simply a group of inspiring entrepreneurs looking to improve customer satisfaction rates and generally make the world a better, more *efficient* place.

Oddly, however, despite not really "doing" politics, Amazon manages to somehow spend over nine million dollars a year on lobbying the government in order to protect its business interests and benefit its bottom line. Specifically, Amazon has helped to lead the fight against internet companies having to pay sales tax, battled against corporate tax reform, attempted to protect its investments in the defence industry, and has continued its "aggressive negotiations" to hamper workforce unionisation.

Amazon has also been embroiled in politics in more direct ways, largely as a result of the firm's cloud hosting business, Amazon

Web Services (AWS). For several years, AWS hosted the popular whistleblowing site Wikileaks. However, following the controversial leak of government documents by Edward Snowden (which highlighted evidence of mass government surveillance and allegations of state-sponsored torture), Amazon took the decision to banish Wikileaks from its servers. As a result, Wikileaks immediately collapsed. In the words of commentator Eli Pariser, "there was nowhere else to go".[17]

For all its libertarian rhetoric, Amazon does "do" politics. The internet is not some magical safe haven which miraculously exists outside the realms of corporate interests and political consequence. Virtual space is simply an extension of physical space and, as a result, must abide by the same laws, standards and expectations of its offline alternative. Removing traditional bureaucrats and gatekeepers does not "free" information, it simply enhances the power of those remaining mediators who have come to understand how best to survive in the new unregulated system.

As a result, we end up with a libertarian society where bureaucratic institutions and Big Government have been banished, only to be replaced with a new less accountable governing force. The American journalist and senior fellow of the New America Foundation, Barry C. Lynn probably described this phenomenon best in a 2014 interview with Salon magazine. According to Lynn:

> "If you hack down government, then you'll have government re-emerge, only the new government will be privately run.... Amazon now essentially governs business within the book industry. Amazon has so much power that it virtually gets to tell really big companies like Hachette, the French publisher, what to do. You're gonna sell this book at this price. You're gonna sell that book at

that price. That means Amazon pretty much has the power to determine how many copies of a book a publisher might sell. That's not citizens trading with one another in an open market setting those prices, that's a giant corporation setting those prices. Which means what we are witnessing in the U.S. book industry, I think, is a form of top-down government."[27]

Yet despite the increasingly top-down role that these businesses play in managing information online, in the rhetoric of a technoutopia, their growing power is generally considered an inevitable part of the march towards progress. Ironically, the actual, democratically elected government is afforded no such luxury online.

When elected governments attempt to pass laws or provide guidance in order to maintain order online, everything they do is scrutinised as an attack on internet freedoms. Even laws designed to prohibit the distribution of child pornography – which would be passed without question in the "real" world – result in a drawn-out debate about internet freedoms.

Governments are typically demonised by the online community, with such demonization often being supported by companies such as Google and Facebook. Despite their democratic mandate to power, much of Silicon Valley still portrays the government as an internet bogeyman, coming to legislate away their freedom through boring bureaucracies. It rarely occurs to these pundits that the internet was not only created by the government, but that most of the best technologies surrounding it, were actually developed through academic research sponsored and paid for by government taxes. That's not even to mention the multi-million-pound R&D grants that the government pays out to companies like Amazon, Google and Apple to ensure that their products "just work".

Of course, such inconvenient truths do not fit with the free market rhetoric of the libertarian internet. That viewpoint rarely makes it into the pages of technology magazines, which are often too busy celebrating the make-or-break courage of "go at it alone" entrepreneurs.

But while we are all celebrating the cult of individuality online, our new heroes are rapidly becoming the very middlemen that the internet was supposed to have cast out. Suddenly, notions of "internet freedom" are increasingly starting to resemble a total lack of corporate accountability. And it is not just the biggest players who are at fault.

In 2013, the lowcost taxi app Uber was embroiled in a San Francisco law dispute after one of its drivers knocked down and killed a six-year-old girl. Given that Uber's drivers were – at the time – not considered official taxi drivers but rather freelance contractors who just happened to have installed the app, Uber denied responsibility for the incident. As far as they were concerned, the driver, and the girl's family, were on their own. Of course, if Uber were an offline company rather than an app, the idea that its employees are not really employees (and so do not merit the same protection rights) would be unlikely to withstand much scrutiny in court. As an app, however, the case took nearly two years to play out, with Uber eventually agreeing to pay a confidential settlement.[28]

This is a running issue that has so far applied to a number of app-based internet start-ups. The holiday rental service Airbnb was embroiled in a similar incident in 2013 following the preventable death of an apartment guest.

As with Uber, Airbnb's stance was that individual hosts were not employees and as such were on their own when it came to liability. This was particularly problematic given that most homeowner's insurance policies will not cover accidents that occur as a result of commercial activities.

Airbnb has also received widespread criticism as a result of its lack of smoke alarms and carbon monoxide detectors. While these would be considered standard safety requirements for any of its offline competitors, Airbnb has previously taken the stance that they do not own nor run any of the properties listed on their site – they are simply an app. Thankfully, as the service has grown in popularity, government pressure and legislation have helped to encourage Airbnb to change its stance, now offering basic liability coverage and a prerequisite requirement for all hosts to install carbon monoxide detectors.[4]

In each of these cases, internet technologies have been heralded as "digital disruptors", cutting out middlemen, breaking barriers to entry and placing power back into the hands of the end customer. Amazon took power back from the publishers, Uber broke the black cab monopoly, Airbnb made travel affordable again. That is how the story goes. But did they really "cut out the middlemen" or have they simply killed the competition and set themselves up as a brand new generation of unaccountable mediators? Has power moved downwards from the gatekeepers to the public, or simply shifted horizontally from the hands of one group of gatekeepers to another?

However the dynamic has changed, for all their faults the likes of Uber, Airbnb and even Amazon still pale in comparison to one group of increasingly unaccountable internet gatekeepers: The search engines.

Search engines: Organising information, on your behalf

As sociologist Alexander Halavais writes, in the internet age, search engines "have become an object of faith".[29] We ask them what to do, what to say, how to think. They are the oracles of our generation, providing reassurance, affirmation and guiding our daily decisions in an increasingly information overloaded world. At the heart of this new religion is Google, the ultimate gatekeeper for the supposedly disintermediated world.

The old cliché that knowledge is power is just as true online as it was in the offline world. The only difference is that, in the internet era, it is not those who create knowledge that hold the power, but rather those that control access to it. Despite all of Silicon Valley's efforts to eliminate traditional gatekeepers, it is ironic that internet technologies have created a generation of possibly the most powerful gatekeepers the world has ever seen – and at their heart, lies Google search.

Currently controlling around 65% of all global search traffic (a figure that jumps to 90% in many European countries), Google now receives over 3.5 billion search queries a day.[30] Looking at my own search statistics, I personally question Google just under a hundred times a day – a usage that does not appear to deviate too far from that of the average tech-savvy consumer. Given this staggering reliance on not only a single corporation, but also a single product, it is hard to disagree with Halavais' view that Google has become less a search engine and more an object of faith. In jest, I have often wondered if even the most devoutly religious ask their god for guidance quite that many times a day.

Ironically, for a technology designed to improve access to information, with every increase in the size and complexity of Google's search database, the number of sources we use to gather information appears to shrink. The more answers Google can provide, the less likely we are to explore alternative information sources. Google not only has the answer, Google *is* the answer.

Turning to this one-size-fits-all information portal has become as natural to us as calling upon the synapses of our own brain, with the verb to "just Google it" now becoming an essential part of the English vocabulary. The more we rely on Google for what, and how to think, the less opportunity we give to those sources – and viewpoints – which do not conform to the mainstream consensus. While an overreliance on Google has proved instrumental in limiting our range of potential sources, this process has further been advanced by our reluctance to explore beyond the first ten results that Google declares as the "best" on offer. Whatever that means.

According to a 2013 study by the online advertising network Chitika, only 6% of searchers ever stray beyond the first page of Google's search results.[31] This means that 94% of users are relying entirely on Google's secretive algorithm to handpick the ten search results – out of billions of potential options – that they will get to see.

For those of us in the marketing community, this comes as little surprise. For years, businesses have struggled and fought to keep their organisations on the first page of Google's prestigious search results, hiring search engine optimisation (SEO) firms in the knowledge that, if they slip off page one of Google, they may as well not exist at all.

Given both the corporate and societal significance of these top ten results, the decision of who or what gets displayed on the first page of Google is a vitally important one. And yet, it is a decision that Google makes independently and with very limited transparency. Due to the severe importance of these search rankings, any suggestion of bias or manipulation of the results must be treated as an extremely serious allegation, one that Google has worked hard to address.

As discussed in a previous chapter, in an effort to uncover the corporate manipulation of search results, Google has launched a variety of algorithm updates designed to weed out potential "blackhat" SEO activities. At the same time, the search firm has also increased both the complexity and secrecy of its central algorithm, with Matt Cutts, head of web spam at Google claiming that the organisation now uses over 200 ranking factors with up to 50 variations per factor.[32] As a result, even the ethical "whitehat" manipulation of search results has become an extremely difficult undertaking.

Whereas Google used to actively encourage the use of SEO to help its algorithm find and effectively crawl a website, many believe that the search giant is finally turning against the SEO industry. Rather than relying on traditional search optimisation techniques, Google now encourages marketers to refocus their efforts on developing genuinely high-quality web content and improved customer experiences. As the search team at Google sees it, if the company's algorithm truly works, people should no longer need to pay for SEO. Instead, the best results should naturally find their way to the top of SERPs (Search Engine Results Pages).

Rather than focusing on how to manipulate the rankings, Google wants marketers and web developers to focus on writing genuinely useful copy and building great web experiences for their customers. Through this free market one-upmanship, it hopes to inspire a race to the top, ultimately resulting in a better, more user-friendly internet for everyone involved.

While this system has proved highly effective in bolstering quality and weeding out biases, limited transparency makes such standards impossible to apply to Google itself. As Geert Lovink argues in his critic of social media, *Networks without a Cause*, "the dangers of filters [are] their invisibility. What we need is a growing awareness of the existence and architecture of the filters that surround us".[33] If we do not know, or are not allowed to know how Google is "organising the world's information", how can we possibly know whether or not the way information is being ordered is fair? If Google was to artificially or unjustly move a website to the top of the results, how would we know about it? More importantly, if a website were to be moved to page nine million of Google (essentially out of public reach), why should the search giant not have to declare why such a decision was made? As long as Google refuses to reveal the "correct" way to organise information, how can we possibly develop a meaningful point of reference for the incorrect way?

For those at Google, secrecy is considered a necessary form of intellectual property protection. Possibly the most common piece of rhetoric used to justify the company's lack of transparency is an analogy between Google's algorithm and the exact recipe for Coca-Cola. While Coke is happy to publish a list of ingredients on every can, to reveal the exact measurements of each ingredient would be to destroy the competitive advantage of the Coca-Cola brand. As a result, it is rumoured that only two Coca-Cola employees may know the full ingredients of the drink at any one time, with the only official written copy (supposedly) held in a top secret US bank vault.

Google treats its algorithm in a similar way, claiming that attempts to reveal the exact details would not only make it easier for people to game the system, but would also allow competitors such as Yahoo and Bing to provide identical results. Google's algorithm *is* its business. On this basis, the search giant has been able to justify keeping the specifics of how it "organises the internet" tightly under wraps.

While it is hard to argue against the idea that Google's algorithm should be part of its intellectual property, for many, the issue of who controls the world's data is too important to be circumvented on such grounds. Given Google's unprecedented power over information access, the protection of Google's intellectual property does not rank as highly as understanding why certain pieces of information are promoted while others are hidden in the darkest depths of Google's search results. It could be argued that such a decision is above any one unelected corporation, let alone one that refuses to reveal its specific decision making process. As Geert Lovink writes, "we need to distrust Google's intention to "organise the world's information" and see it instead as a worldwide move towards data manipulation".[33]

While there are many who would dismiss Lovink's views as an exaggeration of the problem, the mass control of information access – whether by a corporation or otherwise – must be seen as a significant cultural issue to be discussed. As our reliance upon Google grows ever more ubiquitous, those of us in the technology space have come to realise the naivety of our quest to eliminate the gatekeepers of the traditional media. Now, as ever more academics, technologists, and even governments have come to question the monopolistic role of Google in controlling information access, the potential for internal manipulation has moved from a theoretical question to a genuine legal concern.

Calling Google to account

The general lack of transparency surrounding Google's search algorithm makes it virtually impossible to prove the existence of any specific biases within the company's search results. However, as the principles of search neutrality (the idea that search engines should not have editorial policies beyond their algorithms) has evolved within the academic space, several of Google's competitors have begun to wonder whether their products are receiving the correct exposure within the site's seemingly neutral search results.

Many of the most common complaints made with regards to search neutrality have been due to the prominence of Google's own products and services within its search rankings. For some of these products, the decision to place them at the top of

Google's search results seems genuinely justified as they are significantly more popular than their nearest competitors. An example of this would be YouTube, or its sister site Google Videos, both of which appear as the first results when searching for the keyword "videos". Few people would argue that this represents any sort of bias on the part of Google, but rather that YouTube is by far the most popular video streaming service on the internet. If anything, to offer the top spot to less popular sites such as Vimeo or MegaVideo would probably raise even more suspicions about how Google chooses to organise its results. Another example of such a high-ranking service is Google Maps, which is typically listed at the number one spot for the keyword "maps". While some would argue that older results such as streetmap.co.uk or Ordinance Survey should rank above Google Maps, an examination of each site's global usage figures reveals that Google Maps is massively more popular than any online mapping alternative, ultimately justifying its position at the number one slot.

While these examples may appear innocent enough, there have been other high ranking Google services that are more regularly called into question. One example would be Google's mobile app store Google Playstore, which – despite offering fewer apps than Apple's iTunes store – ranks at number 1 for the key phrase "download apps". More worryingly, the iTunes store barely makes it into the top ten results, even when viewed from Safari or an Apple device. While this does not necessarily suggest any form of malicious search bias on the part of Google, it is an unusual observation to note. Either way, the secretive nature of Google's search algorithm ensures that such allegations would be impossible to verify.

In addition to the core rankings, there have also been a number of more subtle examples of potential search bias, many of which have resulted in registered complaints, controversies, and even legal suits aimed at the company. These were less often to do with the company's specific search results, but rather the decision of Google to include particular products and services as part of the wider search engine experience.

One example would be Google Calculator, a basic calculator script which now exists as part and parcel of the Google search engine. Instead of visiting a specific website such as google.com/calculator, visitors to the search engine can now simply type their equations into the search bar and have the answers appear at the top of the SERPs. Alternatively, users can even type the word "calculator" into Google and have the full embedded calculator tool displayed at the top of their results. The only question is, how does this intuitive service fit with the precepts of search neutrality? How does the inclusion of a Google tool permanently at the top of the SERPs coexist with Google's promise to never artificially bolster its own results?

By automatically including its own calculator at the top of the SERPs – even as part of the search engine's functionality – Google is essentially stripping traffic and advertising revenue from the other freely available calculator sites online. At the top of the official (genuine) search results is online-calculator.com, a website which has existed since 2003 providing a free and intuitive calculator tool very similar to that developed by Google. Below online-calculator.com is the Web 2.0 Free Scientific Calculator, followed by a downloadable calculator app. All of these websites have been listed in the top results based on genuine merit (as distinguished by Google's search algorithm), and yet, above them sits an unofficial result, the Google Calculator. This is somehow not quite an advert yet not quite an organic search result; it is simply a tool developed by Google and then placed at the top of their supposedly unbiased search results. Somehow this particular calculator is deemed better than the 464 million results that come below it. Why? Because it was developed by Google, and therefore must sit at the top.

At this point, many people are probably asking the question, "Who cares? It's just a calculator". But regardless of whether Google is using its search results to promote "just a calculator", or its own customer review service, or the embedded Google Flights tool, the precedent is the same. This is a supposedly neutral gatekeeper placing certain information above others, not based on an independent algorithm or a series of predetermined factors, but instead based on a desire to promote and publicise its own products. This is the real point for concern.

And Google is not just promoting its own products, it is also turning a profit on their artificially enhanced placement in the rankings. I wonder how many people still bother to visit online-calculator.com now that Google Calculator is embedded above it? How many people will bother to scroll down and make that extra click to receive an almost identical service? The same could also be said of popular travel services such as Skyscanner. How many people will be visiting the official top result now that Google Flights is seamlessly embedded above?

Many of these internet businesses are kept afloat purely by their high traffic volumes and the associated advertising revenues that this traffic drives. As such, I wonder how much revenue these sites have lost as a result of Google's decision to place its own products at the top of search results? Surely after spotting this dip in traffic any sensible advertiser would shift their pay per click investment to a more popular site – such as advertising on the Google search results themselves.

By providing preferential placement to its own services, Google has fundamentally destroyed all of its competitors within the calculator market. Google Calculator is rapidly becoming the most popular calculator service on the internet, at least for the 65% of the population allows Google to "organise information" on their behalf.

While the loss of revenue to online-caculator.com may seem like a minor offense on the part of Google, this is just the beginning of a far wider trend. With the search giant now including ever more widgets and embedded services at the top of its default search results, the company is slowly strangling competitor ad revenues across hundreds of different fields. As the journalist Jim Edwards writes on *Business Insider*, Google's competitors are not "complaining that Google dominates search" but rather that it "is using that dominance to distort markets outside of the search industry".[34]

This is not the first time that Google has been accused of using its SERPs to interfere in markets outside of search. In 2014, the online review site Yelp accused Google of promoting its own Google Reviews above those of Yelp and Tripadvisor.[34] As part of the company's social networking service Google+, sites that had received positive Google Reviews were ranked higher than those that received positive reviews on Yelp. Working in conjunction with Tripadvisor, Yelp developed a Chrome extension, designed to strip out all search results that were artificially added as a result of Google+. This extension allowed the company to identify that Google was siphoning off 20% of clicks from Yelp's reviews and redirecting them through to Google+ reviews. Once again, Google's reviews were not placed higher up the rankings because they were specifically better than Yelp's, but were instead considered more worthy of the top spot simply because they came from Google. If anything, Yelp and Tripadvisor reviews could be seen as more reliable as they came from an established review platform, rather than from random comments left on Google+.

And it's not just reviews that are being manipulated to the advantage of Google's own products and services. In 2013 the US Federal Trade Commission (FTC) launched a full-scale 160-page report investigating the possibility that Google was using its search results in anticompetitive or unfair ways. While the FTC ultimately decided against pursuing an antitrust lawsuit against the search giant, its initial report found "evidence that Mountain View was demoting its competitors and placing its own services on top of search results lists, even if they weren't as helpful". Specifically "the company used a special algorithm to boost its products' placement on the results page. That algorithm determined which Google product to promote based on what's relevant to the query."[35] The conclusion, according to Engadget's Mariella Moon, was that Google's search algorithm "really was biased."[35]

Scraping the content barrel

Google's decision to promote its own products above "official" search results may have proved controversial within the technology community, but its impact seems inconsequential when compared to the company's use of third party content in promoting its own results.

When displaying information about a particular website or search result, Google will often lift a few lines of content, or "snippet", to be displayed alongside the listing. These snippets are used to provide information instantly and will often include anything from a news headline, photograph, introductory paragraph, embedded video clip or all of the above combined.

While originally developed to populate the search giant's news aggregator, Google News, they can now be found spread across virtually all of Google's various search-related products and services. As one example, a search for John F Kennedy (or any famous individual) will display the standard list of results alongside an interactive snippet taken directly from Wikipedia. This snippet includes a photograph of JFK, a short background of his life, a list of his family members and a series of memorable quotes. Beneath this snippet is a small blue link which credits Wikipedia for providing the information.

In some instances, web designers themselves will actively encourage the inclusion of these snippets as a way to further promote their content within Google. By tweaking the HTML code of their websites, businesses, news sites and blogs can highlight specific areas of content that they would like Google to pick up on. Common examples of these rich snippets of information include five-star restaurant reviews, ecommerce product photos, embedded videos or author information. All of these will then be displayed alongside their websites in the Google search results, boosting engagement and providing their particular site with valuable additional space outside of the plain text Google search results.

Rich snippets may have become a vital part of any website's promotional strategy, but increasingly their unrestrained use by Google has begun to blur the lines between acceptable use and the necessary boundaries of content ownership online. In many cases, rather than waiting for websites to highlight their own content through rich snippets, Google will instead crawl these sites, scrapping and uploading any information that it sees as a useful addition to search results. This is not only done without the content provider's permission, but is also often done without their knowledge.

As is probably to be expected, this unsolicited repurposing of content has proved controversial among content creators and publishers. The first examples of such controversy came about in 2006 following the official roll out of Google News. Despite being ranked as one of the largest news providers in the world, Google News does not actually produce any original content; it is simply a news aggregator, taking a snapshot overview of the content provided by others (headline, blurb, lead image) and ranking it in an easy-to-search format.

While in the days of paid content this could have proved an extremely valuable service for content providers and professional news sites, on the ad-funded internet, Google's use of news snippets has proved yet another unwelcome blow to professional journalists and the wider publishing industry. By providing a list of headlines free of charge, Google News has been accused of siphoning visits away from the front pages of traditional news websites, limiting clicks, and ultimately redirecting advertising revenues away from content generators and towards Google's in-built advertising platform Google Ads. After all, why would anybody bother to visit the Telegraph homepage if all of its key headlines were available via Google News?

Of course, supporters of the search giant argue that Google News can be highly beneficial to traditional publishers, as it increases the availability of their individual news stories online. This, however, does not take into account the fact that the vast majority of website visits are not to individual stories, but rather to a site's homepage. It is this page that has been most damaged by the advent of Google News. This damage has led to a number of high-profile (if often unsuccessful) lawsuits by publishers and content providers looking to reclaim revenue lost as a result of Google News.

And it's not just Google News that has used snippets without permission from the original content provider. Increasingly, direct questions typed into the Google search box will now lift a piece of information from the "best" result and display it at the very top of the SERPs. As one example, a search for "How do I download YouTube videos" not only provides the official results, but has lifted and embedded a step-by-step guide to downloading YouTube videos at the very top of the results. This guide was not written by Google, but was instead lifted directly from ClipGrab.org. Such practices open up a number of serious questions.

Firstly, how has Google decided that ClipGrab.org is the most reliable source of information? It cannot be based on the firm's traditional "organic" algorithm as it is not the top results; in fact, it has not even made it into the top three results. Secondly, and potentially more worryingly, how many people are going to bother to click on the link to ClipGrab.org now that its most valuable piece of information is listed directly above for everyone to see?

The way Google sees it, this is just another way to provide a great and efficient user experience. For ClipGrab.com, however, this is a loss in clicks and therefore advertising revenue that could ultimately result in its business shutting down. Ironically, once taken offline, Google would lose the original snippet and would have to find another website to leech from. It is literally cannibalising its own content.

This is true for all of Google's unregistered snippets. From the "also starred in" list next to actors' names, right through to embedded recipes, thesaurus synonyms and dictionary definitions. Every one of these listings represents stolen clicks and lost revenue from sites such as Wikipedia, IMDB, BBC Good Food and Thesaurus.com. All of which are the very sites that are keeping Google's content current, relevant and alive.

Worst of all, when such sites do attempt to fight Google's unpaid use of content, the company responds with the same mob-shakedown mentality that we have come to expect from Amazon, Verizon and other Silicon Valley giants. In 2006, Belgian copyright management company Copiepresse pursued a claim against Google for including excerpts of its content in Google News. Following a lengthy legal battle, the court told Google that it would have to pay Copiepresse a small fee if it wanted to continue using the firm's content. Instead of paying the fee, however, Google responded by immediately deleting all of Copiepresse' newspapers and client content from its search results. As these sites' traffic ratings plummeted, Copiepresse lost the support of its clients and ultimately had to drop the fee request, pleading with Google to be reinstated in the search results.[36] This has become the pattern with any publishers or journalists that have attempted to regain control over how Google chooses to use their content online, with Google increasingly booting publishers out of search results rather than dealing with the real issue of content ownership.

Probably the most well-publicised instance of this was Spain's now infamous "Google Tax". Following a number of lawsuits against the company for publishing snippets without payment or consent, Spanish politicians introduced a new law that would charge aggregators for using third-party content to generate advertising revenues. The idea (if admittedly poorly implemented) was to ensure that those who developed and wrote content online would at least be getting some cut from the advertising revenues of companies such as Google that were leeching off their content through aggregation. The law was unprecedented and was generally taken as good news for publishers and journalists. Unfortunately, however, the legislation's success or failure could never truly be tested as Google immediately shut down Google News across the whole of Spain.[37] This not only helped to turn Spanish consumers and internet users against the law (and the publishers that had pushed for its enactment), but also sent out a clear message to other publishers around the world: Google will not pay for content, and Google will not negotiate. Or so we thought.

It turns out that those within Silicon Valley have one rule for their own kind and a different rule for everybody else. In mid-2015, Google announced that it would now be including real-time tweets as embedded snippets throughout its core Google Search results. But, unlike the content of lowly journalists, publishers or small-time websites, before proceeding, Google struck an official deal with Twitter in order to outline how and when content can be shown without stepping on any Silicon Valley toes. As part of this deal, Twitter would receive "data-licensing revenue", of around $41 million.[38] As technology journalist Chris O'Brien describes it, "Google is going to pay Twitter for better access to its content".[38]

This represents a massive U-turn (or at least a double standard) from an organisation that has spent the last ten years trying to convince judges, publishers and the wider public that aggregation of content should be free of charge. Suddenly, when a fellow Silicon Valley giant (with the power and financial resources to really tackle Google in court) comes knocking, Google cannot wait to start handing out cheques.

So, to return to the comments of journalist Chris O'Brien... "Where's my cheque?".

Lifting the Google Lens

In terms of Google's role as a new age gatekeeper, aggregation of third party content may have received some of the most vocal criticism, but in many ways, it is merely the tip of the iceberg.

By taking on the task of "organising the world's information", Google as an unelected entity has become the ultimate gatekeeper, far more powerful than any stuffy academic, super rich news mogul, or prudish Mary Whitehouse figure. By deciding how information should be organised on our behalf, Google is not only stripping the irrelevant from view, but is also removing the opportunity to decide for ourselves what is and is not relevant.

That said, however, nobody is being forced to use Google. The search giant does not operate as a benevolent dictatorship, but rather as a company building products so unfathomably excellent that consumers would have to be mad to shun them for a lesser alternative. If internet users really wanted to ditch Google, all of the world's information is still out there online; it just comes down to whether or not anybody could really be bothered to wade through and access it. Theoretically, we could all switch to a different search engine, or even return to the days of tediously typing URLs directly into a web browser, but these notions are easier said than done.

As someone who has tried to give up Google, I know for a fact how difficult it can be to change to a new form of information structuring. A few years ago, following Google's embroilment in various tax-avoidance schemes and the NSA hacking scandals, I decided to make the switch to a more ethical alternative – a search engine that genuinely would "do no evil". Having read a few articles in Wired magazine, I had become aware of a new privacy-conscious search service called DuckDuckGo.

According to its website, DuckDuckGo is a new kind of search engine that emphasises the protection of searchers privacy and guarantees not to track, log or collect any information about its users or their search habits.[39]

After being initially impressed by the site's commitment to individual privacy, as well as its clean user interface (something which many search engines still lack), I must admit that it took me less than four hours to revert back to Google.

That is not to say that there was anything wrong with DuckDuckGo's search results, or even the search experience. If anything, the idea of having an alternative selection of information seemed quite refreshing. Unfortunately, regardless of whether the information was right or wrong, my experiment with DuckDuckGo was doomed from the start, if only because *it wasn't Google*. At this point in time, Google's influence on our daily lives has grown so ubiquitous that any alternative seems inexplicably uncomfortable. The closest analogy I could provide would be to say that I imagine using DuckDuckGo is a bit like driving somebody else's car. The mirrors are at funny angles, the backrest of the chair isn't quite right, and frankly you can't wait to go back to driving your own. We have grown so accustomed to both searching for and receiving information in the way Google has chosen to structure it, that now any alternative to that structure is, by definition, wrong.

Most worrying of all, however, as Google invests in an ever greater number of new technologies, it becomes increasingly difficult to escape the company's particular brand of information organisation. For the so-called "digital natives" – those that were born after the invention of the World Wide Web – it is difficult to remember what the world was like before this single information architecture, or even to remember that alternative models of organisation ever existed.

Inevitably, as the Google brand name has extended into everything from our mobile devices, to our clothing, to our cars, the Google Search algorithm has been built in as the standard model for information retrieval and display.

In addition to Google's 65% share of the web search engine market, Google Search is now included within the main URL bar on Google Chrome – which currently controls around 50% of the overall browser market.[40] In turn, Google Chrome is built in as the default browser on all Android smartphones and tablets – which currently make up over 85% of the mobile market.[40] Android mobiles also incorporate Google as the main knowledge bank behind their voice-activated "personal assistants". These assistants are then set to be included as standard within Google's wearable technology projects, as well as in Google's upcoming driverless cars.

As a result of this ubiquity in the technology marketplace, Google's choices over how information can and should be organised are helping to redefine the way we think about the world. The decisions made by Google's algorithm, no matter how mechanical or unbiased, are creating a filter through which we learn about our surroundings. We see the world through a Google lens, ordering information sources and teaching us which results we should trust and which we should not even have the opportunity to perceive. Worst of all though, as Google becomes the universal source for all personal education, we start to forget that these filters even exist. As Hogeschool van Amsterdam research professor Geert Lovink reminds, the danger of these filters "is their invisibility".[33] Already we are reaching a point where Google has shifted, at least in the eyes of the average user, from a man-made algorithm to an infallible oracle; our very own Pythia for the internet age. What we now need, warns Lovink, is an opportunity to take a step back and develop a new awareness and understanding of the filters that surround us.

Unfortunately, as Google's algorithm has grown more intuitive, the opportunities to develop such an awareness have become increasingly hard to find.

As advancements in individual tracking and personalisation have evolved, Google's decisions of how and when to filter certain information have grown increasingly controversial – further securing the company's role as a modern day gatekeeper. In an effort to serve up the most relevant results, Google personalises its search results based on everything from user browsing histories right through to social networking profiles. As a result of this personalisation, no two searches will ever display the exact same results and no two searchers will ever have access to the exact same information.

As an example, if Google recognises a user's IP address as being based in the US, a search for "football" will display personalised results relating to American football. For searchers in the UK however, "football" would be translated as English football (what Americans would call soccer). This location-based personalisation is not just exclusive to nationality, but will even be applied on a city-by-city, or even street-by-street basis. As such, if a user based in Wimbledon were to search for "Pizza Express", they would receive customised search results relating specifically to the Pizza Express in Wimbledon rather than the main Pizza Express homepage. Similarly, if a user's search history were to reveal that they regularly search for Pizza Express in Clapham (despite living in Wimbledon), or that they just so happened to like "Pizza Express Clapham" on Facebook, Google would use this information to customise the search results further and ensure that details of both stores were available in the SERPS.

This is even true of vague social connections that users have made online. If one of my Facebook friends were to have written a review of a particular Pizza Express on TripAdvisor, that location could appear far higher in my results than it would have in a traditional search. This is supposedly meant to emulate the way real-life friends would recommend restaurants and bars in the real world.

While personalisation has proved an effective way of organising information, and has certainly helped to speed up average search times, it too represents a concerning trend in the way we have come to interact with information.

By customising individuals' results based on their past behaviours, Google walks a fine line between personalisation and stereotyping. Through a complex combination of geographic location, social connections, personal histories and browsing preferences, Google literally makes a judgement on what information users should and should not see as a direct result of who they are and where they come from. Given that the internet is heralded as a hub of information equality, as a form of editorial gatekeeping it is hard to imagine a system that could prove more potentially restrictive.

This idea of personalised information restriction has widely become known as the "filter bubble effect", a term coined by Eli Pariser in the aforementioned book of the same name. According to Pariser, as increasing amounts of our daily lives are controlled and altered by big data analytics and personalisation algorithms, we grow ever more vulnerable to becoming trapped in our own narcissistic bubbles, with our computer monitors and mobile phone screens acting as a "kind of one-way mirror, reflecting your own interests while algorithmic observers watch what you click". [17]

As endless volumes of psychological research can attest, human beings are very good at seeking out the information that reinforces their own views, while subconsciously ignoring that which does not conform. This confirmation bias is common enough without the possibility of handy algorithms purposefully stripping alternative information from view. As an example of this process, Pariser turns to his own Facebook feed which, like Google, is now personalised based on individual histories and previous use:

> "I noticed that my conservative friends had disappeared from my Facebook page. Politically, I lean to the left, but I like to hear what conservatives are thinking, and I've gone out of my way to befriend a few and add them as Facebook connections. I wanted to see what links they'd post, read their comments, and learn a bit from them.
>
> "But their links never turned up in my Top News feed. Facebook was apparently doing the math and noticing that I was still clicking my progressive friends' links more than my conservative friends'."[17]

A similar experience is now happening with each and every one of our Google results, with the search algorithm tailoring our information to place the sites and views that we already agree with or approve of at the very top of the list.

At its most basic level this could simply be a point of concern for those worried about simple stereotyping. Who is to say that a British citizen searching for "football" shouldn't receive results related to American football? Anyone could be interested in American football, that's just geographic pigeonholing. At a more serious level however, personalisation technologies could be seen as limiting the opportunities available to individuals as a result of anything from their income to their tastes in music.

As an example, somebody who lists their previous employment on LinkedIn as a till-worker at McDonalds is unlikely to receive tailored Google Ads promoting higher paid jobs in the city.

Similarly, someone who lists their education at a local state school on Facebook is far less likely to receive advertising or search results promoting Oxford or Cambridge than someone who has listed their private education online. The result is what Pariser jokingly calls "an internet history you're doomed to repeat",[17] a new form of cultural, social and financial stereotyping, in which everyone is boiled down to a series of repeat purchases and algorithmically generated nodding heads.

Opening the gates

According to the pundits of technoutopianism, the internet has proved a vital driving force in the removal of bureaucrats and gatekeepers to universal knowledge. To them, information wants to be free. It does not want to be shackled in the classrooms of elitist universities, collected and analysed by "proper" journalists, or held in the hands of those who believe that content should be paid for or knowledge rewarded. The internet has given way to informational equality, and yet, as we have seen, in a technoutopia some information is more equal than others.

It turns out that removing the supposed gatekeepers of the offline world isn't enough. Toppling journalists and undermining newspaper sales does not create an equality of valuable information, it creates an absence.

In the same way, providing every citizen with a camera phone does not make every citizen a journalist. Journalism, by which I mean "proper" journalism, takes inquiry, investment and training. Such a consideration may not play well with the anti-elitist rhetoric of the modern internet, but it is nevertheless a fact of life.

In the name of efficiency, the internet has replaced professional journalists with bloggers, professional investors with crowdfunding services, and professional teachers with Massively Open Online Courses, all in the name of improved efficiency and an unrealistic vision for equality. Unfortunately, in spite of all the efforts made to bring empowerment and equality to the world, it turns out that the arguments for information equality were circular all along. Through a libertarian consensus, opening up the market to a free exchange of services and ideas did not remove the gatekeepers to knowledge and power, but simply replaced them.

To return to the thoughts of Barry C. Lynn, "If you hack down government, then you'll have government re-emerge, only the new government will be privately run".[27] As the free-market economics of so-called digital disruption have edged their way closer to monopoly, we have ended up with an internet of oligarchs; a surface level of democratic equality being conducted on channels owned and controlled by a handful of Silicon Valley billionaires.

The ubiquity of these channels ensures that our information

access is both restricted and perverted through helpful personalisation. Top secret algorithms ensure that views that do not reinforce our own are cast to the bottom of the pile, while others are not even displayed to us at all. Individuals' content is lifted and stolen under the mantra that information should be free, while insider deals and aggressive lawsuits ensure the strict copyrights of Silicon Valley's "Masters of the Universe" are not impeached upon. Those that do not like it can always move to another service, or they could be forcibly removed from their existing service. Outside of social and search, businesses such as Uber and Airbnb "digitally disrupt" the market by failing to protect their drivers or fit basic safety precautions, while Amazon uses its monopolistic stranglehold over the ecommerce market to disappear competitor products over night. To the marketing departments of Silicon Valley, this doesn't represent a lack of basic customer protection but rather a "transformative change".

In the background, the technoutopians remind us that algorithms and apps are not capable of bias, while the vast lobbying teams of Facebook and Google tell us that they "don't do politics".

We live in the age of the invisible gatekeeper. Never before have we had such a vast platform on which to share our thoughts, and yet, never before have we delegated so much of our thinking to a handful of unelected elites.

To return to the thoughts of writer Nicholas Carr, "in an automated system, power concentrates with those who control the programming".[7]

The internet improves our ability to communicate and socialise

"Sociable technology will always disappoint because it promises what it cannot deliver. It promises friendship but can only deliver performances."

Sherry Turkle

If there is one thing the internet can be universally praised for, it is the ability to connect people. From Skype, to Facebook, to Twitter, to Tinder, to Chatroulette, the internet lets friends, families and even strangers communicate, socialise and connect at real-time speeds from anywhere around the world. It allows people to contact their relatives thousands of miles away, lets users view photos and videos from events they never had a chance to experience, and even allows us to build new and exciting relationships with people we would have once never even met. At its heart the inter*net* is exactly that, a network, designed to allow communication between individuals all over the world. Socialisation and human communication are not just part of the internet; they are the internet.

Marshal McLuhan once claimed that new media would ultimately turn the world into a global village – an "extension of consciousness" in which cross-world communication would allow all of the earth's peoples and cultures to blend into a single homogeneous society; an international community with the close-knit mentality of a local village.[1]

Of course, writing in 1962, McLuhan had no idea how close to reality his notions of "electronic independence" would later become. Thankfully the likes of Bill Gates were around to complete his analogy, adding that the internet will become "the town square for the global village of tomorrow".[2]

Given the widespread impact of internet communication upon all of our lives, it's easy to understand why such sentiments have been lapped up by the internet community. The idea of a global village somehow makes the internet seem quite quaint, polite and fragile. We can nip across the village green to the local library (Wikipedia), grab two tickets at the cinema (Netflix), or pop into the coffee shop on the corner (Facebook) to catch up with old friends.

If anything, the global village seems closer to a 1990s vision of the internet, with strangers sharing advice on Usenet groups and long lost friends catching up on email. On the modern web however, somehow the internet's image no longer seems as quaint.

If the internet is a village, it is one that has been sold off and privatised by large multinationals. CCTV cameras litter the streets. Friends shout over one another for attention. The library contains a baffling array of facts, half-truths and political propaganda. Occasionally, a flasher will open his trench coat on the lawn, before being lucidly "flagged for review" by some unseen, unelected, private police force.

For those versed in the rhetoric of technoutopia, to complain about such issues is simply to dwell on the inevitable. For them, technology remains a tool to be wielded. If someone chooses to use their camcorder to make a sex tape instead of a wholesome home movie, does that put the camcorder at fault? Of course not.

While such arguments remain convincing on the surface, they force us to narrow our critical scope to a series of drastically simplified deterministic inevitabilities. This deterministic rhetoric should be treated with the same caution as the libertarian argument that "guns don't kill people; people do". The internet is not some tool to be wielded, nor is it an unstoppable force which can never be legislated or controlled. The internet is a highly complex social, cultural and political environment, well deserving of discussion, study – and yes – critical analysis.

The internet may very well be the greatest achievement in human communication since the development of language, but this however is not a reason to bolster it above the realms of criticism – quite the opposite. It is the fact that the internet is such an important and all-encompassing communications medium that makes it so deserving of critical review. When examining a technology as obviously beneficial as the internet, the technorealist approach is more vital than ever. It is needed to ensure that we suspend our awe and do not judge such tools on their achievements alone.

To return to another of Marshal McLuhan's famous adages, "the medium is the message". With the introduction of every new medium, be it a platform, a channel or even a new device, the content being served will always be perverted to a new interpretation. Books naturally interpreted content as imagination, television interpreted it as entertainment, and now computers interpret the world as information to be processed. As the American architect, Richard Saul Wurman writes in *Information Anxiety*, "the form in which information is presented will dramatically alter the perceptions of that information. It will also change the very nature of the information itself".[3]

In the age of social media, the "message" is not a television program, novel, or even an article of news, it is human interaction itself. How we communicate and interact with one another as fellow human beings is now more than ever being conducted through the filter of the internet. Given the importance of this subject, and the sheer volume of internet communication that occurs every day, is it not wise to take a step back and consider how the message of human communication has been altered by the medium of the internet? Is it not worth asking whether communication has grown more widespread or whether we have simply expanded its very definition? Does the internet provide better communication or simply more communication? Does Facebook allow us to have more friends than ever before, or does it simply stretch the definition of a "friend" beyond traditional recognition? Do Twitter and Chatroulette allow us to bask in the company of friendly strangers, or simply glimpse at the constructed masks of people who crave anonymity and personal redefinition?

These are the questions that the technorealist paradigm demands we ask, both of the internet and of our own socialisation and communication habits. Unfortunately, these are all questions that are becoming increasingly difficult to answer.

As digital natives become the dominant demographic in our society, very soon, few will even remember how communication used to occur before the age of constant tweets, texts and self-obsessed selfies. For many young people, online friends are considered just as valid as real-life relationships. Likewise, swiping right on Tinder is increasingly seen as the only springhead of potential romances and sexual explorations.

As the digital natives of today become the boring, stuffy adults of tomorrow, one has to wonder how much human interaction and communication will change. Will the presented self of Facebook, Instagram and Twitter become the real self? Will carefully filtered and corporate-monitored communication become the only kind we really know?

As Neil Postman wrote back in 1985, "to be unaware that a technology comes equipped with a program for social change, to maintain that technology is neutral, to make the assumption that technology is always a friend to culture, at this late hour, is stupidity plain and simple."[4] It is this cynical yet necessarily critical approach that is undertaken throughout the rest of this chapter.

For the hundreds of books, magazine articles and blogs that sing the praises of the internet's communication revolution, too few bother to ask what aspects of communications culture are lost in the process of technological change. It is these losses that will be examined going forward; losses on a social, a cultural and a highly individualistic level. For as Neil Postman concludes:

> "Every new technology... involves a trade-off. It giveth and taketh away, although not quite in equal measure. Media change does not necessarily result in equilibrium. It sometimes creates more than it destroys. Sometimes, it is the other way around." [4]

The online world is flat. You are flat.

To a computer, everything is information, and in the information age, everything is flat.

Computers deal in numbers and data, data which can be processed logically and then repurposed as a problem to ultimately be solved. The remaining qualitative subtleties and nuances that do not easily translate are either abandoned, repurposed, or simply ignored. Ideas such as culture and interpretation are increasingly viewed as fluffy and outdated, as relics from a time before scientific quantification and big data analysis reigned supreme.

Social media is no different, it is just the product that has changed. If anything, the product being processed on sites such as Facebook and Instagram is even more complex and nuanced than anything that has previously been transferred online, it is our interactions, our personalities, our very selves.

To say that the internet flattens everything is not merely a comment on the limitations of technology, it is a statement on the unfettered optimism of those who develop the code. For all the media portrayals of Mark Zuckerberg as an architype of the autistic genius trope, there may be some truth to this overly stereotyped narrative. For Zuckerberg and his fellow believers, Facebook is not a two-dimensional representation of people's lives, it is an accurate catalogue. To them, the idea of converting complex and qualitative aspects of life into a series of tick boxes, dropdown menus and analytics graphs is simply a reflection of the fact that people aren't really that complicated. They do not have multi-faceted personalities, and even if they did, that is simply evidence of their weak social characters. As Zuckerberg himself has previously claimed, people only have "one identity", having "two identities for yourself is an example of a lack of integrity".[5]

It is difficult to say if this aggressive simplification of human nature is down to genuine utopianism, a personal lack of social understanding, or is simply a well-rehearsed lie designed to convince Facebook advertisers that their audience data is remotely meaningful outside of the platform. Whatever the reasoning behind it, it is a mentality that appears to be spreading.

As citizens have made way for netizens and netizens have made way for information-consumers, simplification has shifted from an undesirable offshoot of laziness to a widely expected benefit of modern existence. The phrase "there's an app for that" is more than just a marketing slogan for the new generation of information consumers, it is a credo through which they demand to live their lives. While consumerist society has certainly helped in cementing this expectation, internet technologies have also turned this passive expectation into an active demand. Everything in our world must be presented as a simple query to be searched, an app to be installed, or a logical problem to be instantly solved – all on demand, at our fingertips and free of charge.

If we are lost, we do not risk exploration, but rather convert our problem into a query to be processed by Google Maps. If we are hungry we do not experiment with the ingredients available, but rather turn to the star-ranked recipes of BBC Good Food, or lazily tap open the JustEat app on our iPads. Everything in our lives is becoming simple, so why should human interaction be any different?

As social media, and specifically social networking technologies, have brought us together online, this expectation of quantifiable simplicity has leaked into our social lives and our day-to-day interactions. Unfortunately, unlike the process of finding a hotel or ordering a takeaway, human interaction is rarely as simple as it first appears.

Across the various fields of psychology, communication theory, linguistics and anthropology, academics and researchers have uncovered barely the tip of the iceberg when it comes to the nuances of human communication.

From cultural variations, to non-verbal signs, to body language, to group-mentalities, to tactile changes in temperature, human communication is about so much more than just a status update. To accept this fact however is to give in to the idea that simplification and efficiency may not always be the correct path to choose – an admission which flies in the face of the internet's mantra of quantification as the ultimate source of information equality.

As a result, the optimistic developers of Web 2.0 set out to tame the social animal; to create online "communities" where everyone could speak and share their thoughts openly. Information consumers were quick to jump on-board, flattening themselves and their personalities for easy digestion. And it's fully understandable why they would want to do so. There is something extremely satisfying about the process of flattening your personality into a series of quantifiable traits. It is like you are getting to know yourself better, coming to terms with your various attributes, pointing at them and saying "There it is. That right there, is me."

Building a social media profile provides a quick and appealing journey of self-discovery, without the need to climb a mountain, visit a monastery, or go sit in the woods for two months with Henry David Thoreau. The fact that we haven't actually discovered anything new, or even developed an accurate portrayal of ourselves, is irrelevant. The illusion of self-mastery remains the same.

As rational beings we crave a coherent narrative, yet our lives remain one of the least coherent narratives it is possible to imagine. We despise flip-floppers and hypocrisy, yet must permanently alter our opinions in order to learn and grow. We see those who change their personalities to fit in as weak-willed, yet within almost every social group we find ourselves offering some degree of personal compromise. We want to be perceived as intelligent and funny, yet in the vast majority of cases our intellect falls short and our wit arrives several hours after the necessary set up has occurred.

Clearly, Zuckerberg's view that holding multiple personalities is a sign of limited integrity is little more than the opinion of a man who does not understand human communication. As entrepreneur Andrew Keen clarifies, "having multiple identities – as a citizen, a friend, a worker, a woman, a parent, an online buddy – is actually an example of somebody with so much integrity that they are unable to compromise on their different social roles".[6]

Everything about the Facebook platform betrays this lack of understanding, right down to Zuckerberg's own description of the site. Rather than listing Facebook among other social networks, Zuckerberg prefers to reference it as a "social utility", alongside water companies and electricity providers.[7] While it is easy to understand why Facebook may want to position itself alongside these essential services, it is hard to think of a worse way to describe a platform designed to mimic human communication. To label such a complex, qualitative and abstract process as little more than a utility reeks of the cold, simplified scientism that has come to define the current phase of the social internet.

Through a maze of dropdown menus and status updates, technoutopians have set out to define the human character, to neatly categorise each of us into our own coherent narratives. Worst of all, we could not wait to let them do so.

Sites like Facebook have allowed us to remove the complexities of social interaction, along with the complexities of our own contradictory selves. By shifting to the online realm we have redefined the boundaries and definitions of human socialisation. To be "friends" with someone is now an entirely new game. We no longer have the option to lose touch. Nor do we have the option to learn or grow, we have been, to use the psychologist Sherry Turkle's phrase "flattened into personae", reduced "to a list of favourite things".[8]

This list of "things" is admittedly expanding, with Facebook continuing to add a baffling array of historical geo-locations, gender options, relationship statuses, likes and moods. But each of these options will continue to exist in isolation, failing to interact within the meaningful contexts needed to understand a fully-fledged human.

Facebook allows us to craft and ultimately judge our peers on a series of checkboxes and simplistic forms. As Sherry Turkle explains in *Alone Together*, "in real life, people can see you are cool even if you like some uncool things. In a profile, there is no room for error. You are reduced to a series of right and wrong choices."[8]

These "right or wrong choices" may seem like a bit of harmless fun, but they are increasingly coming to define us as individuals across all walks of modern life. In the early 2000s, employers would have had to have been mad to have "stalked" their employees on Second Life, or to have monitored their phone calls for comments against the company culture. Likewise, insurance firms felt no realistic need to monitor individual forums for mentions of regular alcohol consumption. The reason? People knew that what was being said and shared online probably wasn't a reliable depiction of how these individuals led their offline lives.

As social networking has grown ubiquitous and technoutopians have come to honestly see their products as an adequate alternative to traditional interaction, our carefully crafted (if wholly inaccurate) online selves have become the sign posts for who we really are.

The only problem with this shift, however, is that it fails to account for the possibility that genuine people are genuinely complicated – a fact that organisations like Facebook purposely avoid. Facebook does not want to know the real you as the real you is messy and difficult to quantify. At the end of the day, no advertiser is going to want to buy a big messy real-life person; as such, you have to be simplified.

Increasingly, social media services (particularly 'profile sites' such as Facebook) are presenting a reductionist view of personal identity. They bolster their revenues by promoting the idea that all people consist of a series of quantitative facts to be stored and downloaded into a .csv file. As a result, sites like Facebook often assume that the world is not complicated, or that if it is, it should be filed away under an 'It's complicated' dropdown selection. If they ever were to admit, as Upworthy founder Eli Pariser once did, that there is "no one set of data that describes who we are",[9] they would undermine the entire data-driven advertising model upon which the modern web is based. Worse still, it would undermine the technoutopian notion that everything can be quantified and that all aspects of life can be converted into raw information

#Awkward

Possibly the biggest change that the internet, and in particular social media, has brought to the process of human interaction is the ability to pause and think; to buffer in silence before writing and rewriting our response. In many ways, this is a positive addition to the typically frantic and reactionary communication style. Where face-to-face and voice communication forced us into instant and ill-thought-through responses, now, the seemingly natural barricade of an online interface provides us time to process each piece of information, absorbing it fully in order to develop the best possible response. Through this socially acceptable buffering, our online personas can become better communicators than our real-life selves. We can become more thoughtful, more rational, and even more sensitive to the comments and needs of others. Best of all, we can improve (or at least pretend to improve) our outward intelligence, reducing our chances of stumbling blindly into a conversation we know little to nothing about. The days of social awkwardness are behind us. Now our communication is polished to perfection, Googled, spell-checked, written and rewritten to convey exactly what we would have said – had we only been more intellectually prepared or a genuinely wittier person.

As internet communication has become the dominant form of interaction within the Western world, this barrier of slow-thinking pseudo-intellectualism has provided a comfortable shield, without which we would all feel overly exposed. This is particularly true of the younger generation, which has abandoned telephone and video communication in favour of instant messaging, texts and one-way social media posts. In fact, many studies suggest that young people are actually afraid to use their phones for anything other than text messaging and internet use, finding the very process of talking without the ability to reword or buffer, extremely intimidating.[10] As Sherry Turkle summarises, "our networked life allows us to hide from each other, even as we are tethered to each other... we'd rather text than talk".[8]

This is the irony of online communication. Through its mediated nature it allows us to hide from our socially awkward real-time selves, and yet, in removing the need to practice real-time communication, it makes our remaining "real life" social interactions more awkward than ever before.

At the same time, by living our social lives through the shield of mediated communication, we risk removing the opportunities for genuine dialogue, understanding and forgiveness that socially uncomfortable face-to-face communication would have produced.

As an example, consider the process of "unfriending" someone on Facebook – the complete removal of a person's access to your profile, timeline, updates, instant messaging and video communication across the site. In the "flesh world" – to use William Gibson's superbly nineties term – the process of unfriending someone is a slow, painful and enormously awkward affair. You could ignore them at work, or let their calls ring to voicemail. You could try to avoid them in the supermarket, or stop attending their parties. Sooner or later though, you were bound to bump into them and be forced to interact. Were this chance meeting to occur, you had one of two choices: either you could blank them and ignore the problem, or you could discuss the issue and "have it out" with them.

With regards to the first option, remaining silent gives us the opportunity to reflect internally on why we are angry, upset or disappointed. It also allows us to examine why we did not choose to confront the issue with them when we had the chance. Perhaps is it not as a big a deal as we originally thought? Perhaps we do not want to discuss it because deep down we know we are being petty, vindictive, or just plain in the wrong.

Considering the second option, "having it out" affords us the opportunity to discuss our grievances. It provides our friends with the opportunity to apologise or argue their own case. In some instances it may even allow us to uncover unknown circumstances that could drastically change our point of view.

Whichever of these physical paths we choose to take, they both share a common benefit in that they provide us with the time and opportunity to think, reflect, rationalise, discuss, and ultimately, forgive. Online communication is a far less forgiving arena. Temporary arguments, disagreements or even imagined slights can now be "solved" through a simple unfollow, unfriend or even a block button. Now, if you do not like somebody, you simply do not *like* them. If you do not want to see them anymore, you can unfollow or unfriend them. If you disagree with what they have to say you can stick them on mute or block them from your profile. By applying such black and white yes/no, friend/foe filters to human communication and companionship we are drastically simplifying the process of social bonding. By removing the social awkwardness involved in unfriending someone, we remove the opportunity to forgive and learn from our grievances. Of course, there is always the option that we may forget the original slight and come to unblock or re-friend our colleagues further down the line. Forgetting, however, is no substitute for forgiveness, and our relationships will inevitably suffer as a result.

The same is also true of our relationships with our partners. Since its introduction to Facebook in the early 2000s, the online relationship status has become a staple part of social networking.

The ability to define oneself as single, in a relationship, or the often ironically used, "it's complicated" represented a drastic shift in way teenagers and young adults negotiated their love lives online. Suddenly to be "official" was not enough, in order to be taken seriously as a couple you needed to be "Facebook official". For Facebook users of a certain age this notion quickly became a conventional milestone of relationship development – even extending to the offline realm. As Samuel Axon writes on Mashable, "changing Facebook relationship status has, for better or worse, joined first date, first kiss, first night together, exclusivity talk, and first "I love you" on the list of important relationship milestones".[11]

In the same way that Facebook changed what it meant to be "friends" or to "unfriend" someone, the incorporation of online relationship statuses has altered the process of relationship development and even relationship collapse. Where once, ending a relationship used to be a reasonably private and unofficial thing, now the need to publicly update one's profile welcomes a torrent of complications, questions and often unwelcomed sympathy. As with the process of unfriending a person, changing relationship status at the click of a button does not provide the same opportunities to rethink or reassess. By making something Facebook official we have stripped the social awkwardness needed to learn from our decisions.

While I cannot imagine that many mature adults are breaking up via a change in Facebook relationship status (although apparently it does happen), there is still a lost opportunity for potential resolution. By switching between the binary dropdowns of 'single' and 'in a relationship', we are forced to commit our decision to the public record, making it all the more difficult to undo at a later date. Nobody likes to be seen as uncertain in their decisions, nor do they enjoy the social embarrassment that comes from perceived hypocrisy or flip-flopping. As a result, the process of endorsing or announcing something in a public space leaves little wiggle room to reassess or perform a 180-degree turn.

The impact of such public announcements has been clearly demonstrated in the fields of consumer and marketing psychology. It is one of the reasons why marketers and brands are so keen for you to "like" their products online – partially because it provides social proof to your friends and family that the product is trustworthy, but also because once you have endorsed a product online you are committed to that decision.[12] Changing your mind becomes far more challenging as it means publicly admitting that the 'like-happy' you of the past was wrong and is capable of suffering a lapse in judgement. It is the same reason why those who join public weight loss groups are typically more successful in their goals than those who try to quietly diet and exercise on their own. To announce your endeavour publicly is to provide yourself with the opportunity to fail in front of your peers, as such there is a far greater incentive to remain resolute and ultimately succeed.

While this public commitment can prove beneficial in striving for weight loss, quitting smoking, or any other self-improvement programme, when it comes to improved communication or relationships, such public pressures do little more than remove the opportunity to internally reflect. This places the fear of public embarrassment above the desire to do what could be genuinely right for you. As the journalist and editor Michael Harris asserts in his critique of the connected world, *The End of Absence*, when we post our ideas and decisions online:

> "we abandon the powerful workshop of the lone mind, where we puzzle through the mysteries of our own existence without reference to the demands of an often ruthless public. Our ideas wilt when exposed to scrutiny too early – and that includes our ideas about

ourselves".[13]

Once again, the narrowing of our relationships is an example of supposedly social technologies taking the slow, complex and often uncomfortable processes of human interaction and attempting to flatten them onto a more comfortable, if weaker, binary landscape.

As social networking sites have solidified our decisions and our pasts for public display, the internet users of today have started to construct ever more complex walls around their online selves. What once was an opportunity to share our real-time thoughts and construct an accurate personal profile is now an opportunity to rebrand ourselves as something better: a more consistent, confident and intelligent self. Social networking has become an opportunity to construct one's personality with the same precision and careful planning of a marketer building a brand; an opportunity to share the right content, support the right causes and portray the right ideals. This is the age of personality as a product, it is the age of Brand Me.

Brand Me

> "The overall goal with branding is to differentiate yourself (the product) in the market so you can attain your objectives, be those landing your dream job or becoming a famous singer. The process includes defining your brand and brand attributes, positioning your brand in a different way than your competitors and then managing all aspects of your personal brand".[14]

Those depressing words were recently written by Lisa Quast, in an article titled *Personal Branding 101* for Forbes Magazine.

Quast describes herself as a career coach who sets out to "maximise people's potential",[14] a vague yet important task which, based on her various Forbes articles, I am sure she is very good at.

At the heart of this personal improvement lies a commitment to turn her subjects into "personal brands", a task that she is not alone in. All over the internet, social media gurus and self-improvement professionals are jumping on the *Brand Me* bandwagon, reminding people that everything they do online is a representation of their personal brands. And they're right. Thanks to the unfading memory of sites such as Facebook and Google, the information that individuals post online is inescapable and will be tied to their "brands" forever. While ego-driven marketers see this as an opportunity to build some kind of thought-leadership role for themselves, for most of us this should not represent a healthy method of personality formation.

In some ways, this cult of narcissism does not find its roots in technology, but rather in the rise of celebrity.

Twenty-five years ago, a psychological study was conducted asking children in the UK which jobs they would most like to see themselves eventually going into. The results were telling of the ambitions and priorities of the time; with teacher, banker and doctor being the three most commonly chosen professions for five to eleven-year-olds. In 2009, the study was repeated. This time however the most common career selection for five to eleven-year-olds was footballer, closely followed by pop star and then actor. In short, the young people of today want nothing more than to be celebrities; to stand at the centre of the stage with the spotlight and the flashbulbs directly on them. Is it any wonder that the rhetoric of personal branding resonates so well with the new generation of social media stars?

As appalling as the phrase "personal brand" first appears, there is something particularly apt about it. Brands are, after all, nonauthentic. They are fairytales, spun by marketers to help sell products on the back of a fictitious idea – that products can change your personality. Through the power of branding, we have become convinced that products somehow stand for anything other than what is inside the tin. Apparently Dove soap stands for women's rights and the campaign for realistic standards of female beauty. Similarly, Lynx deodorant is supposed to stand for raw sex appeal and macho domination over women. The fact that these brands are both owned and produced by Unilever is irrelevant. The story they tell is all that really matters.

The same is true of personal branding. Brands are, by their nature, inauthentic, and that is exactly what your personality must become in order to succeed as a personal brand. Social media did not start this trend, but it has fueled the fires of self-obsession throughout today's youth. In the early days of Web 2.0, people started blogs and GeoCities pages as a way to catalogue and showcase their lives. While it seems fair to assume that a lot of these bloggers would have painted their lives in a favourable light, generally speaking the purpose of web logging was always to provide an honest and authentic portrayal of one's life in the real world. If anything, people were more authentic online, using the anonymity provided by forums and message groups to share the ideas and concerns that they did not feel comfortable expressing in the real world.

Now, the opposite is true. Social media actively encourages its users to "rebrand" in a positive light. Facebook adds effects to photographs, Snapchat adds filters to make people thin, Vine adds music in order to dramatize real world situations. Of course, we all know that these additions are fake, but we do not seem to care. We have reached a stage where teenagers are happy to take selfies with their friends and then, without even thinking, apply a Snapchat effect to "touch up" the image and make themselves look better. Once, photographs were used to capture a memory, a lasting image of a genuine experience or event. Now however, we have reached a stage where it is acceptable to airbrush your friends in a single click; to fake a memory without a second thought.

In some ways, the use of filters and quirky photo effects can be seen as little more than a form of play. If, however, that is the case, then social media should not claim to capture our lives or truly represent our personal identities online. Facebook Timeline cannot and should not be a lifelong record of falsified memories, but that is increasingly what it has become. Through a combination of careful curation, selective uploading, and yes, even bare-faced airbrushing, social media has become less of a communications medium and more of a supermarket shelf full of carefully marketed products and brands.

This is not to say that there are no benefits to becoming a brand. If anything, the ability to project a clear consistent image will almost certainly have its advantages when developing and maintaining a strong social identity.

For an example, we need only consider the rise of the heated online political debate. While people have long debated politics with their friends and family in the real world, most of us have a sensible cut off point at which a healthy discussion turns into a full-blown argument. Usually, it is at this point that we know when to hold our tongues. Within the online realm however, something about the safety of a mediated user interface makes us forget the social norms that would traditionally contain us. On my own Facebook feed I have seen countless friends and relatives virtually screaming at each other across the comments section of a political post. Similarly, I have seen some (usually older) relatives sharing things that they would never dream of saying in a physical public space. Strangely, while supposedly allowing people the opportunity to buffer and think, social interfaces seem to encourage people to unleash levels of certainty, self-confidence and abuse that they would never dream of resorting to within the real world. As internet critic Geert Lovink explains, "the internet is a breeding ground for extreme opinions and border-testing users… The public internet has turned into a battlezone, which explains the success of walled gardens such as Facebook".[15]

The irony, of course, is that for every self-confident political assertion we make online, the more we have to adapt our personal brands in order to maintain a consistent narrative. As our confidence grows and our personal assertions amplify, the opportunity to reassess or alter our views decreases. It is hard enough to admit internally that we are wrong, but once we have committed our seemingly fixed views to writing and have shared them in front of our colleagues and friends, it becomes almost impossible to change our views. Our ill-thought-out assertions become part of our personality, they become part of our personal brand.

Perhaps the result of this will eventually be that we become coyer in our online communication? Perhaps the "angry internet" is little more than a passing phase? Writing in *Me and My Web Shadow*, Anthony Mayfield introduces the idea of a personal web "footprint", an online shadow which needs to be managed everywhere we go online.[16] As the original internet generation, perhaps we are just not yet accustomed to managing this footprint. Perhaps future generations will not resort to any loud assertions online for fear that they may permanently damage their personal brands. This could be the true irony of the *Brand Me* age; that in an era when everyone believes they have something to say, the eternal nature of the internet itself may be the one thing that stops them from saying it.

Downloading selfishness

While the social buffer of network interfaces has clearly encouraged people to assert their views online, this explosion of confidence has not been helped by the self-reinforcing nature of internet technologies themselves. As discussed in the previous chapter, Eli Pariser's filter bubble effect is not only impacting our ability to access alternative views, it is also limiting our opportunities for genuine discourse and varied communication online. With each bold statement that we make on social media, the automated algorithms of modern personalisation are encouraged to recommend content that reinforces our own views. For those on the left of politics, Facebook will bump up content that supports their pre-existing opinions, the same is also true for those on the right.

While this continuous reinforcement may appear gratifying and even useful when attempting to win an argument, it does not provide a healthy foundation upon which rational and measured communication can be built. In a supposedly open forum such as social media, people need to feel able to change their minds. More importantly, they also need to be provided with information that offers an alternative view. That is the only logical basis for a healthy debate.

And it is not just debate that is suffering under the strain of increasingly targeted personalization; individual identities are also at risk.

Having accepted the notion that we are all curating our own personal brands, the rise of personalisation makes it increasingly difficult for us to experiment with how we want these brands (our identities) to be positioned and received. With every targeted post, personalised result or piece of recommended reading, we close down unexplored avenues for personal development. Our personalities grow stagnant, permanently faced with content and ideas that reinforce a view of ourselves dictated by an earlier time. As Pariser explains "while the internet can give us new opportunities to grow and experiment with our identities, the economics of personalisation push towards a static conception of personhood".[9]

Having promised us a new era of social communication and personal choice, what the internet has actually delivered is quite the opposite. The days of communicating with strangers and experimenting with our identities online have been replaced with a system of increasingly strict monism of the self. There can only be one me, and that me is continuously self-reinforcing.

As our notion of self grows stronger, and a desire for consistency cements increasingly volatile opinions within our personal brands, the desire to compromise and empathise with our fellow human beings wears thin. Selfishness is typically the follow-on from self-absorption and self-absorption more often than not finds its roots in self-confidence. We grow accustomed to being reinforced and to being right and that inevitably leads to a shift in the way we perceive our importance and the importance of our fellow citizens. This self-absorption is not helped by the callus nature of our new social interfaces. What started with the narrowing down of relationships – the ability to unfriend and unlike people with the click of a mouse – has spiraled into something even less emotionally appealing. Through the gamification of social interaction, sites like Tinder, Grindr and Chatroutlette encourage users to only value physical appearance, literally swiping people out of their lives based on nothing more than a nonchalant glance at their face. Tinder has become the ultimate meat market, a place where human interaction, sexual desire and even love can be boiled down to a shallow, Darwinist fight for the best piece of prime rib.

On a brief side note, this metaphor for Tinder was brilliantly satirised by the Dutch artists, Cors Brinkman, Jeroen van

Oorschot, Marcello Maureira, and Matei Szabo. They launched a Tinder account entirely controlled by a single, spinning piece of chopped liver. As far as I am aware, the so-called 'Tender' project is still live and can be seen swiping people right to this day.

Meat metaphors aside, there is a serious point to be made about the way in which 'Tinderfication' is changing how we make and sustain connections online. By creating a semi-physical, semi-anonymous barricade between us and our digital peers, we open ourselves up to a new wave of objectification and personal selfishness. To return to the thoughts of Sherry Turkle, "the connected life encourages us to treat those we meet online in something of the same way we treat objects."[8] People online are not quite artificial, but they are also not quite people. As a result, the way we treat them does not accurately reflect the way that we would behave – or expect others to behave – in the offline world.

It is this convergence between the real and the unreal that has allowed meaningful communication and civilised debate to crumble online. For the new generation, social technologies do not represent mutual understanding, collective discussion or some quaint Global Village in which we all get along. They represent faux outrage, cyberbullying, online trolling, negative commentary and a new era of vicious and ill-researched intellectual spam.

Worst of all, as this new form of selfish, hyper-enraged discourse becomes the standard for communication online, there is a strong argument to be made that such changes in human interaction may find themselves slipping out of the ethereal and into the real world.

The technologies that have proved so damaging to online communication are now making their way into our physical lives. Already, the ability to put people "on pause" has seeped into the acceptable norms of communication behaviour. How many times have you been in the middle of a conversation with someone when they have opted to take a phone call? Lifting a finger to your face, they physically pause your presence while jumping to another conversation, like a Facebook users hopping between chat boxes and ignoring your reply.

The integration of digital technology into the physical world is also set to blur the lines further and make this issue significantly worse. Already, advances in communication technology have allowed the 'Tinderfication' of love to spread into the offline dating scene, with the combination of algorithmic dating and Google Glass proving a particularly creepy vision for the future of romance. One app, NameTag, can be used to snap photos of potential mates and cross reference them in real-time with social media, dating profiles and even the sex offenders' register.[17] Supposedly, the vision behind this app is to take the risk out of dating, helping singletons find their perfect match without the social embarrassment of approaching people that they may not get along with. Of course, where such a notion immediately falls down is in the typically technoutopian assumption that quantifiable data should lie at the heart of every decision. The possibility that serendipity still has a role to play, or that two people with completely different backgrounds – and browsing histories – could get along has become unthinkable.

Another "advancement" in real world communication inspired by social media has been the development of a physical mute button – allowing people to literally "block" certain sounds, or even voices, that they do not want to hear. Reminiscent of Charlie Brooker's bleak television series *Black Mirror*, the Hear Active mute system has so far received over $600,000 in donations on Kickstarter.[18] Incorporating the latest active noise cancelling technology, Hear Active can be used to block out specific frequencies including crying babies, background noise, or simply the entire world around you. While it is easy to see the potential demand for such a device, many commentators are concerned that this system could be used to break down communication entirely. As with the block button on Facebook, a real life muting system cuts down the opportunities for people to learn, listen and ultimately forgive each other for past grievances. Looking ahead, such audio curation technology could one day provide people with the opportunity to personalise their environment in the same way they would personalise their social media feeds, cutting out those voices and opinions that prove disagreeable or personally challenging. As tech culture journalist Christopher Hooton explains, this system "could be a handy way of blocking out inequality. A more nuanced version might allow users to selectively mute protests for instance, or poor people, or entire cities and communities using geo-location".[19] The possibilities are both endless and terrifying.

An unconformable place to live

According to the tenets of technoutopia, the internet provides us with the greatest communications tool ever developed – a fact that is hard to deny. At the heart of the modern internet lies social media, the beating pulse of digital communication that allows us to socialise with strangers, keep in touch with old friends and even find new loved ones online. Every second nearly 8000 tweets are sent, 1000 Tumblr posts are written, 800 Instagram photos are shared and 2000 Skype calls are made, culminating in more social interactions than ever before.[20] For the technorealist however, the question remains whether these interactions are genuine, and more importantly whether they should be treated as simple additions to our current communications arsenal or whether they are set to replace the existing norms of social interaction altogether.

To return to the writings of Neil Postman, "every new technology… involves a trade-off. It giveth and taketh away, although not quite in equal measure".[4] If there is any trade-off worthy of slow and considered study, it is that of communication. Communication lies at the heart of humanity, informing knowledge, morality, politics and culture. In ensuring the sanctity of these things, a reliance on technoutopianism should not be considered enough.

As this chapter has explained, social media and digital communications have slowly altered the way that we as individuals perceive both ourselves and one another. The ability to buffer, research and craft has allowed us to curate our conversations, building responses that are simultaneously more accurate and yet somehow less genuine. With each assertion made, our views have been amplified and enforced through recommended reading lists and self-indulgent personalisation algorithms. Having never taken the time – or having never had the opportunity – to explore the alternative point of view, our reactions grow more volatile and we hold stronger and louder beliefs. Our comments on social media grow crueler, we correct friends and argue with relatives over meaningless squabbles that we once would have quietly let slide. With each one-sided argument we make, our conviction grows stronger. The social awkwardness of being proved wrong in front of our peers makes it increasingly hard to back down. Instead, we simply adapt these new views into our personas and come to believe that we have always been "that sort of person". Having flattened ourselves to fit within the limited dropdown menus of social networking sites, we must then curate every response, every post and every share to reflect some sort of consistent narrative. The possibility that human identities are fluid and inconsistent by nature is not an acceptable view to hold.

As our personal brands solidify and the nuances of interaction wear down to fit the limitations of social media, the very compromises that define internet communication begin to seep into the offline realm. We expect the opportunity to pause and mute, to buffer and research, to always know the right answer all of the time. We also begin to expect consistency, both from ourselves and from those around us. The people who portray such a funny and confident image online are failing to make us laugh in the real world. We are let down by their fallibility, their bumbling apprehensions and their innately boring humanness.

Through the limitations of social media, we have created a world in which everyone can share their thoughts yet nobody is willing to speak their minds. Where anyone can showcase their personalities but everyone is pretending to be someone else. Where memories can be captured forever yet filters and Photoshop are used to distort reality. Where everyone is equal yet everyone is an object to be swiped or a brand to be bought.

In the same way that news media is merely a representation of the news, social media is nothing more than a representation of genuine social interaction. Constant communication may have led to a higher quantity of interactions, but the quality of those interactions is paper thin. The global village that was once envisaged by Marshal McLuhan has become an uncomfortable place to live. Now it is populated by strangers who know nothing about one another other than the most superficial of facts. Led by blind technoutopianism, it has become a connected society without genuine connection. A society in which technology leads, and people follow.

A realist conclusion

Throughout the course of this book we have seen the ways in which internet technologies have subtly impacted our culture and our daily lives. We have witnessed the confusion between access and transparency, the inability of information to ensure equality and the downfall of investigative journalism in favour of clickbait and PR. We have also found numerous examples of monopolisation across the free internet, with traditional gatekeepers being ousted and then replaced with a new generation of Silicon Valley oligarchs. We have seen how information architecture is subtly shaping our worldview, and how the rise of crowd-based wisdom is providing an acceptable platform for the very worst of mob mentality. We have come to understand how man-made algorithms are distorting our information access and how politicians, dictators and corporations can use our own infobesity against us. At a human level, we have witnessed the need for unique individuals to water down their personalities, cramming them into tick boxes, profiles and standardised dropdown lists. Finally, we have witnessed the rise of consumerism online and how it encourages a new generation of teenagers to view themselves as fixed products to be experienced rather than as fallible, ever-evolving human beings.

Combining these factors together into a single list, it would be all too easy to come away from this book with a bleak outlook on technology, and possibly even progress itself. That however, has never been my intention.

When I started writing this book, I selected the name Technoutopia knowing full well that the end result was likely to be something almost entirely different. The aim of this book, however, was never to attack technology, nor to be Luddite or even dystopian in its outlook. This is not a call to arms, nor is it a call for people to unplug their routers, delete their Facebook profiles or go live in the woods. If anything, this book is the opposite of a call to arms, it is a call for nuance. All I ask for is an opportunity to hit pause. An opportunity to collectively buffer and assess the direction in which we are heading.

In an age of lightning fast communication in which major decisions are demanded at equal lightning speed, this collective buffering is more important than ever before. It is also more difficult than ever before to achieve.

In a society that idolises gut-reaction entrepreneurship and the mythology of the twenty-something Wunderkind, those who request time to think are considered at best stuffy academics and at worst tedious bureaucrats. To be anything but certain is to be a flip-flopper, and to question the value of technological "disruption" is to be accused of romanticism and nostalgia.

Such views are not facts, but rather the hardcoded mythology of a society engulfed in technoutopianism. As Neil Postman wrote nearly twenty years ago, we are currently surrounded by "one-eyed prophets who see only what new technologies can do and are incapable of imagining what they will undo. They gaze on technology as a lover does his beloved, seeing it as without blemish and entertaining no apprehension for the future".[1] In the time that has passed since Postman first wrote those words, technoutopianism has spread further and faster than he could have ever imagined.

Today, planning for the cultural implications of a new technology is not only out of fashion, it is positively unthinkable. Thirty years of libertarian rhetoric has ensured that, to those in Silicon Valley, consideration for such consequences is taken as equivalent to fighting the future itself. I, for one, do not subscribe to such a deterministic view, and I would like to hope that this book has provided a fairly substantial list of reasons why you should not accept it either.

Technology and progress are not one and the same thing. Even more importantly, not all progress is destined to be positive. As discussed at the start of this book, we live in an age of swelling populations, depleting resources and the imminent threat of anthropogenic climate change. Our society suffers from an epidemic of mental illness, while online harassment contributes to a thirty-year high in suicide rates. Privacy withers in the face of advertising, while journalism cosies up to PR. In digital commerce, a new generation of monoliths use a combination of predatory pricing and tax avoidance to eliminate the competition, ultimately limiting consumer power. And yet, to the new generation of technoutopians this is all inevitable.

As the electric lightbulb was destined to put the candlestick makers out of work, so too is Amazon destined to put high-street retailers out of business. Similarly, autonomous vehicles are doomed to make taxi drivers unemployed, while robots will almost certainly put the vast majority of manufacturing jobs at risk. Whether or not these things are inevitable, however, is not an excuse to give up on discussing them, questioning them and ultimately managing them through. It's also not a valid reason to stop questioning whether they will be genuinely preferable in the long term. Will an autonomous taxi driver be better than a human chauffer? Will the convenience of Amazon genuinely outweigh the community of a traditional high-street? These are all questions that we still have time to ask; questions which should not be swayed by talk of deterministic progress. This is especially true if the vision for that progress is being shaped and funded by some small handful of Silicon Valley entrepreneurs. If we are to invest in one notion of progress or another, it must be a notion that is collectively and democratically defined by society itself.

This is not to downplay the role that these evangelists and

entrepreneurs have in our lives, nor should it be perceived as an attack specifically on market economics or the nature of capitalism. As an experiment in both libertarianism and free market economics, the internet has *generally* been extremely successful. Where legislation has been implemented it has often proved ham-fisted and near impossible to regulate. Where consumer demand has been allowed to flourish however, market forces have often provided the best end result.

One of the most obvious examples of this was the rise of online piracy. Since the days of Napster and LimeWire, governments and corporations have attempted to use legislation to block the ability of pirates to illegally download music, movies and other electronic content. Through a combination of updated laws, ISP blocks, raids, cease and desist letters, fines, public awareness campaigns and even prison sentences, governments and corporations have tried to discourage the use of BitTorrent and related peer-to-peer sharing networks. While this has certainly made it more difficult to download content illegally, the main outcome of such efforts has been a war of escalation. For every block or law put in place by the government, online pirates have come up with new and increasingly innovative ways to hide their activities online (proxies, VPNs, Tor, dark net streaming services).

Where government legislation failed however, free market forces have proved a far more effective means of limiting piracy online. While many of the largest music and film companies were busy fighting for tougher legislation, online streaming companies such as Spotify and Netflix were considering why so many consumers had turned to piracy in the first place. As they understood it, people were not risking colossal fines because they did not want to pay for content, but rather because the ability to instantly access any content offered a far better user experience than going out and buying a physical album or DVD. Netflix and Spotify realised that the best way to fight music piracy was to create similarly low cost services that provided the same instant access and user experience offered by online piracy. As these companies have grown and their products have become the go-to platforms for online streaming, 2016 saw traditional internet piracy fall to a record low.[2] This was not a triumph for legislation, but rather a clear example of free enterprise solving a problem through a genuine desire for progress.

The purpose of this example is not to undermine the need for legislation online, but rather to emphasise that the point of this book is not to push a legislative agenda. Instead, it is simply a request for governments, businesses, internet users and citizens to slow down and really question how technology is impacting their lives. Services such as Netflix and Spotify may have helped to end online piracy, but they have also created a new wave of problems related to the production and distribution of content online. Since the introduction of Spotify, musicians have become trapped in the brand's increasingly monopolistic approach to content distribution. The result has been a new generation of artists who are haemorrhaging profits at a similar rate to that caused by music piracy in the first place. The internet may have provided a platform through which they can easily disseminate their music, but somehow they have still ended up trapped in the same structure of monopolistic gatekeepers that defined the music industry offline. The same is increasingly true of software distribution, video content and even digital literature.

It is these subtle consequences and complications that I hope this book has helped to establish and to question in the face of rampant technoutopianism. As we have seen, the commonly held tenets of internet optimism are myths. Information freedom is not synonymous with transparency. Access to non-contextual information is not making us smarter by default. Endless social communication channels are not making us inherently more social. Internet access is not a substitute – or even a prerequisite for democracy. Most importantly, any ethereal link between technology and progress is built on a foundation of evangelistic optimism, libertarian rhetoric and utopian mythology. This is not to say that such views are maliciously untrue, but rather that they are simply one overly-idealised side of a much wider story. They are the voice of utopianism in a world that should triumph realism over positivity. The only problem is that developing a realist view takes time – the one thing that an information economy cannot provide.

What we need is a little room to breathe; room to develop a realistic approach to how we can integrate technology into our society and how that in turn should come to define our vision of genuine progress.

Twenty years ago a small group of academics, technologists, cyberpunks and journalists attempted to create such an approach. They were the original technorealists, whose mantra called for a more nuanced method to technological adoption. To them, the internet was not the basis for some glorious alternative system, but rather a simple extension of our own flawed society. As with any other walk of life, managing the internet would take thought, time, cultural consideration and, where required, government legislation. Sadly, such nuanced ideas were ultimately lost in a haze of technutopianism.

What is needed now is a return to realism, a return to uncertainty, to scepticism, and a renewed alertness to the ways in which our creations are influencing our daily lives.

If society is ever to reach for a true utopia, it will not be crafted in HTML or launched as an iPhone app. It will instead be fashioned in the slow and nuanced debates of citizens, activists and elected representatives all across the globe.

References

Introduction

1. Page, H. (1996). Top technology trends on the horizon. Available: https://www.entrepreneur.com/article/13550. Last accessed 23rd October 2016.

2. Shenk, D. (1997). Data Smog: Surviving the Information Glut. New York: HarperCollins Publishers.

3. Shenk, D. et al (1998). Principles of technorealism. Available: http://www.technorealism.org/. Last accessed 23rd October 2016.

4. Shenk, D. et al (1998). Technorealism SIGNUP FORM. Available: http://www.technorealism.org/cgi-bin/cgiwrap/dshenk/db?action=view. Last accessed 23rd October 2016.

5. Georgiadou, l. (2002). McLuhan's Global Village and the Internet. 2nd International Conference on Typography and Visual Communication.

6. Lai, J. et al (2009). Information wants to be free ... and expensive. Available: http://fortune.com/2009/07/20/information-wants-to-be-free-and-expensive/. Last accessed 23rd October 2016.

7. Bemasek, A. & Mongan, D. (2015). All You Can Pay: How companies use our data to empty our wallets. New York: Nation Books.

8. Klein, N.(2008). The Shock Doctrine: The Rise of Disaster Capitalism. New York: Penguin.

9. Postman, N (1985). Amusing ourselves to death. York: Methuen & Co.

10. McLuhan, M. (1964). Understanding Media: The Extension of Man. Reading: Cox & Wyman Ltd.

11. Postman, N. & Powers, S. (2008). How to watch TV news. New York: Penguin.

12. Postman, N. (1992). Technopoly: The Surrender of Culture to Technology. New York: Vintage Books.

13. Dempsey, R. (2013). Online Privacy Never Existed. Available: http://robertwdempsey.com/online-privacy-never-existed/. Last accessed 29th October 2016.

14. Weisman, A. (2013). Countdown: Our Last, Best Hope for a Future on Earth?. London: Little Brown and Company.

15. Popovich, N. (2013). US suicide rate soars to 30-year high in growing epidemic across America. The Guardian. Available: https://www.theguardian.com/us-news/2016/apr/22/us-suicide-rate-30-year-high-growing-epidemic-across-america. Last accessed 29th October 2016.

16. May, G. (2015). Why don't Japanese men like having sex?. The Telegraph. Available: http://www.telegraph.co.uk/men/thinking-

man/11362306/Why-dont-Japanese-men-like-having-sex.html. Last accessed 29th October 2016.

17. Turkle, S. (2013). Alone Together. New York: Basic Books.

The internet encourages democracy, equality, and transparency

1. Powell, L & Shandley, R (2016). German Television: Historical and Theoretical Perspectives. New York: Berghahn Books.

2. Shapiro, A. (1999). The Control Revolution: How the Internet is Putting Individuals in Charge and Changing the World We Know. New York: Public Affairs™.

3. Shirky, C. (2008). Here Comes Everybody: The Power of Organizing without Organizations. London: Penguin Group.

4. Solis, B. (2011). A Social Democracy: The White House Learns To Listen. Available: http://www.briansolis.com/2011/06/a-social-democracy-the-white-house-learns-to-listen/. Last accessed 9th October 2016.

5. Jarvis, JJ. (2010). My cyberspace bill of rights. Available: https://www.theguardian.com/commentisfree/2010/mar/29/internet-censorship-cyberspace-bill-of-rights. Last accessed 9th October 2016.

6. Blake, A. (2016). Defense Distributed, designer of 3-D printable guns, loses appeal against State Dept.. Available: http://www.washingtontimes.com/news/2016/sep/22/defense-distributed-designer-of-3-d-printable-guns/. Last accessed 9th

October 2016.

7. Morozov, E. (2013). To Save Everything Click Here. London: Penguin Group.

8. Shenk, D. (1997). Data Smog: Surviving the Information Glut. New York: HarperCollins Publishers.

9. Sabur, R. (2015). National biscuit day: what does your favourite biscuit say about you?. The Telegraph. Available: http://www.telegraph.co.uk/news/newstopics/howaboutthat/11635912/National-biscuit-day-what-does-your-favourite-biscuit-say-about-you.html. Last accessed 9th October 2016.

10. Hooton, C. (2015). You can now ship your enemies a bag of dicks for them to eat. The Independent. Available: http://www.independent.co.uk/news/weird-news/you-can-now-ship-your-enemies-a-bag-of-dicks-for-them-to-eat-10118997.html. Last accessed 9th October 2016.

11. O'Connor, D. (2009). Witness the Freest Economy: the Internet. Mises Institute. Available: https://mises.org/library/witness-freest-economy-internet. Last accessed 30th October 2016.

12. Chua, A. (2000). The Paradox of Free Market Democracy: Rethinking Development Policy. Harvard International Law journal. 41 (2). Available http://scholarship.law.duke.edu/cgi/viewcontent.cgi?article=5578&context=faculty_scholarship

13. Dahl, R, A. (2000). On democracy. London: Yale University Press.

14. Keen, A. (2014). The internet is not the answer. London: Atlantic Books

15. Liebling, A, J. (1960). Do you belong in journalism. The wayward press. The New Yorker. Available http://www.newyorker.com/magazine/1960/05/14/do-you-belong-in-journalism

16. Shirky, C. (2005). Power Laws, Weblogs and Inequality. In: Lebkowsky, J Extreme Democracy. Carolina: Lulu.com.

17. Shirky, C. (2009). Here Comes Everybody: How Change Happens when People Come Together. London. Penguin Group.

18. Halavais, A. (2008). Search Engine Society. Cambridge. Polity.

19. Lee, J. (2013). No. 1 Position in Google Gets 33% of Search Traffic [Study]. Available: https://searchenginewatch.com/sew/study/2276184/no-1-position-in-google-gets-33-of-search-traffic-study. Last accessed 31st October 2016.

20. Bergman, M. (2001). The Deep Web: Surfacing Hidden Value. Harvard International Law journal. 7 (1). Available http://quod.lib.umich.edu/j/jep/3336451.0007.104?view=text;rgn=main Last accessed 31st October 2016.

21. Zipf, G. (2012). Human Behavior and the Principle of Least Effort: An Introduction to Human Ecology. Eastford: Martino Fine Books.

22. Graves, P. (2010). Consumer.ology: The Market Research Myth, the Truth about Consumers and the Psychology of Shopping. Boston: Nicholas Brealey Publishing.

23. Shankland, S. (2008). We're all guinea pigs in Google's search experiment. CNET. Available https://www.cnet.com/uk/news/were-all-guinea-pigs-in-googles-search-experiment/ Last accessed 31st October 2016.

24. Lutz, A. (2012). These 6 Corporations Control 90% Of The Media In America. Business Insider. Available http://www.businessinsider.com/these-6-corporations-control-90-of-the-media-in-america-2012-6?IR=T Last accessed 31st October 2016.

25. Morozov, E. (2013).To Save Everything, Click Here. London: Penguin.

26. Birkinbine, B. et al (2016). Global media giants. London: Routledge.

27. Edmonds, R. (2016). Newspaper declines accelerate, latest Pew Research finds, other sectors healthier. Available: http://www.poynter.org/2016/newspaper-declines-accelerate-latest-pew-research-finds-other-sectors-healthier/416657/. Last accessed 31st October 2016.

28. Bemasek, A. & Mongan, D. (2015). ALl You Can Pay: How companies use our data to empty our wallets. New York: Nation Books.

29. Chomsky, N. & Herman, E. (1988). Manufacturing consent:

The Political Economy of the Mass Media. New York: Pantheon Books.

30. Cromwell, D. & Edwars, D. (2014). Guardians of Power: The Myth of the Liberal Media. North Carolina: Lulu Press Inc.

31. Kahneman, D. (2012). Thinking, Fast and Slow. London: Penguin.

32. Bold, B. (2014). Top 100 UK advertisers: BSkyB increases lead as P&G, BT and Unilever reduce adspend. Available: http://www.campaignlive.co.uk/article/1289560/top-100-uk-advertisers-bskyb-increases-lead-p-g-bt-unilever-reduce-adspend Last accessed 31st October 2016.

33. Cromwell, D. & Edwars, D. (2009). Newspeak in the 21st century. London: Pluto Press.

34. Aula, P. & Heinonen J. (2015) The Reputable Firm: How Digitalization of Communication Is Revolutionizing Reputation Management. New York: Springer Publishing.

35. Stack, L. (2015). BuzzFeed Says Posts Were Deleted Because of Advertising Pressure. New York Times. Available: http://www.nytimes.com/2015/04/20/business/media/buzzfeed-says-posts-were-deleted-because-of-advertising-pressure.html Last accessed 31st October 2016.

36. BuzzFeed Advertise. (2016) Buzzfeed. Available: https://www.buzzfeed.com/advertise Last accessed 31st October 2016.

37. Crikey. (2010). Over half your news is spin. Crikey. Available:

https://www.crikey.com.au/2010/03/15/over-half-your-news-is-spin/ Last accessed 31st October 2016.

38. Bernays, E. (1928). Propaganda. New York: IG Publishing.

39. Simmons, S. (1972). New speakers handbook: How to be the Life of the Podium. New York: Amacom.

40. Waddington, S. (2016). What is public relations? CIPR. Available: http://influence.cipr.co.uk/2016/05/03/what-is-public-relations/ Last accessed 6th November 2016.

41. Miller, D. & Dinan, W. (2007). A Century of Spin: How Public Relations Became the Cutting Edge of Corporate Power. London: Pluto Press.

41. Stauber, J. & Rampton, S. (2004). Toxic sludge is good for you!. London: Constable & Robinson Ltd.

42. Lee, J. (2013). No. 1 Position in Google Gets 33% of Search Traffic [Study]. Search Engine Watch. Available: https://searchenginewatch.com/sew/study/2276184/no-1-position-in-google-gets-33-of-search-traffic-study Last accessed 6th November 2016.

43. Elliot, N. (2009). The Easiest Way to a First-Page Ranking on Google. Forrester. Available: http://blogs.forrester.com/interactive_marketing/2009/01/the-easiest-way.html Last accessed 6th November 2016.

44. Holiday, R. (2012). Trust Me, I'm Lying. London: Penguin.

45. George Washington University & Cision (2009). 2009 Social

Media & Online Usage Study. Available https://www2.gwu.edu/~newsctr/10/pdfs/gw_cision_sm_study_09.PDF Last accessed 31st October 2016.

46. Morozov, E. (2013).To Save Everything, Click Here. London: Penguin.

47. Kovach, B. & Rosenstiel (2011). Blur: How to Know What's True in the Age of Information Overload. New York: Bloomsbury Publishing PLC.

48. Waddington, S. & Earl, S. (2012). Brand Anarchy: Managing Corporate Reputation. London: A & C Black Publishers Ltd.

49. Montgomery, D. et al. (2013). Police, citizens and technology factor into Boston bombing probe. Washington Post. Available: https://www.washingtonpost.com/world/national-security/inside-the-investigation-of-the-boston-marathon-bombing/2013/04/20/19d8c322-a8ff-11e2-b029-8fb7e977ef71_story.html. Last accessed 29th October 2016.

50. Charles, C. (2015). 3 examples of social media witch hunts gone wrong. Available: http://www.thatsnonsense.com/3-examples-of-social-media-witch-hunts-gone-wrong/. Last accessed 29th October 2016.

51. Madrigal, A. (2013). #BostonBombing: The Anatomy of a Misinformation Disaster. The Atlantic Available: http://www.theatlantic.com/technology/archive/2013/04/-bostonbombing-the-anatomy-of-a-misinformation-disaster/275155/. Last accessed 29th October 2016.

52. Mirkinson, J. (2014). New York Post's Defense Of Notorious

Boston Marathon Cover Gets Laughed Out Of Court. The Huffingston Post. http://www.huffingtonpost.com/2014/03/08/judge-new-york-post-boston-marathon_n_4925255.html. Last accessed 29th October 2016.

53. LoGiurato, B. (2013). FBI Shreds The Media Over Unverified Boston Marathon Reports. Business Insider. http://www.businessinsider.com/boston-marathon-arrest-bombing-reports-fbi-2013-4?IR=T. Last accessed 29th October 2016.

54. Terbush, J. (2013). N.Y. Post under fire for misidentifying Boston bombing 'suspects'. The Week. http://theweek.com/articles/465326/ny-post-under-fire-misidentifying-boston-bombing-suspects. Last accessed 29th October 2016.

55. Ackerman, S. (2013). This Is the Modern Manhunt: The FBI, the Hive Mind and the Boston Bombers. Wired. https://www.wired.com/2013/04/boston-data-manhunt/. Last accessed 11th November 2016.

56. Meyers, P. (2011). Duplicate Content in a Post-Panda World. Moz. https://moz.com/blog/duplicate-content-in-a-post-panda-world. Last accessed 11th November 2016.

57. Adams, O. (2014). Is Google Destroying The Press Release?. Come Recommended. comerecommended.com/google-destroying-press-release/. Last accessed 11th November 2016.

58. Southern, M. (2014). PR Newswire Responds To Panda 4.0 By Taking Action Against Spammers. Search Enine Journal.

https://www.searchenginejournal.com/pr-newswire-responds-panda-4-0-taking-action-spammers/110476/. Last accessed 11th November 2016.

59. Shearman, S. (2014). Panda 4.0: Good news for content, bad news for link-stuffing. PR Week.
http://www.prweek.com/article/1301803/panda-40-good-news-content-bad-news-link-stuffing. Last accessed 11th November 2016.

60. Google. (2016). Link schemes.
https://support.google.com/webmasters/answer/66356?hl=en. Last accessed 11th November 2016.

61. Fishkin, R. (2009). Google's Sandbox Still Exists: Exemplified by Grader.com. Moz.
https://moz.com/blog/googles-sandbox-still-exists-exemplified-by-gradercom. Last accessed 11th November 2016.

62. Baldwin, C. (2013). Google penalises Interflora and UK newspapers for failing to comply with SEO regulations. Computer Weekly.
http://www.computerweekly.com/news/2240178518/Google-penalises-Interflora-and-UK-newspapers-for-failing-to-comply-with-SEO-regulationsm. Last accessed 11th November 2016.

63. Google. (2014). Google Bowling - Google Search.
https://www.google.co.uk/search?q=Google+Bowling&oq=Google+Bowling&aqs=chrome..69i57j69i64.86j0j4&sourceid=chrome&ie=UTF-8. Last accessed 8th September 2014.

64. Waddington, S. (2012). Corporations are people: British Airways. Wadds. http://wadds.co.uk/2012/11/18/corporations-

are-people-british-airways/. Last accessed 11th November 2016.

65. Sourcewatch (2014). Front Groups: Examples. http://www.sourcewatch.org/index.php/Front_groups. Last accessed 11th November 2016.

66. Sourcewatch (2008). Non-Smoker Protection Committee. http://www.sourcewatch.org/index.php/Non-Smoker_Protection_Committee. Last accessed 11th November 2016.

67. Sourcewatch (2016). Independent Women's Forum. http://www.sourcewatch.org/index.php/Independent_Women%27s_Forume. Last accessed 11th November 2016.

68. Kidnetic (2007). Bright Papers: Eating for Energy. http://www.kidnetic.com/brightpapers/?c=Featured&p=44. Last accessed 11th November 2016.

69. Robertson, J. (2014). Peabody Energy targeted by climate activists at G20-related energy forum. The Guardian. https://www.theguardian.com/australia-news/2014/nov/12/peabody-energy-targeted-by-climate-activists-at-g20-related-energy-forum. Last accessed 11th November 2016.

70. Mcnevin, G. (2014). Has Peabody been fibbing about its G20 coal supporters? The Australian. http://www.theaustralian.com.au/business/business-spectator/has-peabody-been-fibbing-about-its-g20-coal-supporters/news-story/8513a47b593f11baa6a2579402cd142c. Last accessed 11th November 2016.

71. Wikipedia (2016). Conflict-of-interest editing on Wikipedia. https://en.wikipedia.org/wiki/Conflict-of-interest_editing_on_Wikipedia. Last accessed 11th November 2016.

72. Craver, J. (2015). PR firm covertly edits the Wikipedia entries of its celebrity clients? Wiki Strategies. https://wikistrategies.net/sunshine-sachs/. Last accessed 11th November 2016.

73. Arthur, C. (2013). Wikipedia sends cease-and-desist letter to PR firm offering paid edits to site. The Guardian. https://www.theguardian.com/technology/2013/nov/21/wikipedia-cease-and-desist-pr-firm-offering-paid-edits. Last accessed 11th November 2016.

74. Pegg, D. (2011). Wikipedia founder attacks Bell Pottinger for 'ethical blindness'. Independent. http://www.independent.co.uk/news/uk/politics/wikipedia-founder-attacks-bell-pottinger-for-ethical-blindness-6273836.html. Last accessed 11th November 2016.

75. Lee, D. (2012). Wikipedia investigates PR firm Bell Pottinger's edits. BBC News. http://www.bbc.co.uk/news/technology-16084861. Last accessed 11th November 2016.

76. Postman, N. (1992). Technopoly: The Surrender of Culture to Technology. New York: Vintage Books.

The internet removes bureaucrats, intermediaries and gatekeepers

1. Finley, K. (2016). Google Fiber Just Swallowed Up Another Internet Provider. Wired. https://www.wired.com/2016/06/google-fiber-just-swallowed-another-internet-provider/. Last accessed 11th November 2016.

2. Shenk, D. (1997). Data Smog: Surviving the Information Glut. New York: HarperCollins Publishers.

3. Keen, A. (2008). The Cult of the Amateur: How blogs, MySpace, YouTube and the rest of today's user-generated media are killing our culture and economy. London: Nicholas Brealey Publishing.

4. Keen, A. (2014). The internet is not the answer. London: Atlantic Books.

5. Wohlsen, M. (2013). For Google's Founders, What's Cooler Than a Private Jet? A Private Terminal. Wired. https://www.wired.com/2013/04/private-jet-terminal-for-google-founders/. Last accessed 11th November 2016.

6. Rushkoff, D. (1999). Coercion: Why we listen to what "they" say. New York: Riverhead Books.

7. Carr, N. (2015). The Glass Cage - Where automation is taking us. London: The Bodley Head.

8. Cohen, N. (2008). He Wrote 200,000 Books (but Computers Did Some of the Work). The New York Times. http://www.nytimes.com/2008/04/14/business/media/14link.html. Last accessed 11th November 2016.

9. Ingram, M. (2014). A print newspaper generated by robots: Is this the future of media or just a sideshow?. Gigaom. https://gigaom.com/2014/04/14/a-print-newspaper-generated-by-robots-is-this-the-future-of-media-or-just-a-sideshow/. Last accessed 11th November 2016.

10. Ingram, L. (2013). Blinkist Book Summaries Arrive to Improve Your Commute And Make You Look Smart. TechCrunch. https://techcrunch.com/2013/11/14/blinkist-book-summaries-arrive-to-improve-your-commute-and-make-you-look-smart/. Last accessed 11th November 2016.

11. Morozov, E. (2012). The Net Delusion: How Not to Liberate The World. London: Penguin.

12. Tovey, A. (2014). Ten million jobs at risk from advancing technology. The Telegraph. http://www.telegraph.co.uk/finance/newsbysector/industry/11219688/Ten-million-jobs-at-risk-from-advancing-technology.html. Last accessed 11th November 2016.

13. Cooper, S. (2012). Instagram's Small Workforce Legitimizes Other Small Start-Ups. Forbes. http://www.forbes.com/sites/stevecooper/2012/04/17/instagrams-small-workforce-legitimizes-other-small-start-ups/#321151f53818. Last accessed 11th November 2016.

14. Burnham, K. (2014). Facebook's WhatsApp Buy: 10 Staggering Stats. InformationWeek. http://www.informationweek.com/software/social/facebooks-whatsapp-buy-10-staggering-stats-/d/d-id/1113927. Last accessed 11th November 2016.

15. Harris, M. (2014). The End of Absence: Reclaiming What We've Lost in a World of Constant Connection. London: Portfolio Penguin.

16. Jenkins, S. (2009). Name, shame, blame the bankers, if you like. But they're the wrong target. Guardian. https://www.theguardian.com/commentisfree/2009/nov/26/name-shame-bankers-wrong-target. Last accessed 11th November 2016.

17. Pariser, E. (2012). The Filter Bubble: What The Internet Is Hiding From You. London: Penguin.

18. Mangalindan, J. (2010). Jeff Bezos's mission: Compelling small publishers to think big. Fortune. http://fortune.com/2010/06/29/jeff-bezoss-mission-compelling-small-publishers-to-think-big/. Last accessed 11th November 2016.

19. Wikipedia. (2016). Amazon.com controversies. https://en.wikipedia.org/wiki/Amazon.com_controversies. Last accessed 11th November 2016.

20. Stone, B. (2014). The Everything Store: Jeff Bezos and the Age of Amazon. London: Corgi.

21. Gessen, K. (2010). The war of the words. Vanity Fair. http://www.vanityfair.com/news/business/2014/12/amazon-hachette-ebook-publishing. Last accessed 11th November 2016.

22. Streitfeld, D. (2010). Plot Thickens as 900 Writers Battle Amazon. The New York Times.

http://www.nytimes.com/2014/08/08/business/media/plot-thickens-as-900-writers-battle-amazon.html. Last accessed 11th November 2016.

23. Hightower, J. (2014). 4 ways Amazon's ruthless practices are crushing local economies. Salon. http://www.salon.com/2014/09/27/4_ways_amazons_ruthless_practices_are_crushing_local_economies_partner/. Last accessed 11th November 2016.

24. Bezos, J. (2008). Jeff Bezos: The King Of E-Commerce. Entrepreneur. https://www.entrepreneur.com/article/197608. Last accessed 11th November 2016.

25. Soper, S. (2015). Amazon to Ban Sale of Apple, Google Video-Streaming Devices. Bloomberg. https://www.bloomberg.com/news/articles/2015-10-01/amazon-will-ban-sale-of-apple-google-video-streaming-devices. Last accessed 11th November 2016.

26. Brad, S. (2009). Amazon Erases Orwell Books From Kindle. New York Times. http://www.nytimes.com/2009/07/18/technology/companies/18amazon.html. Last accessed 11th November 2016.

27. Frank, T. (2014). Free markets killed capitalism: Ayn Rand, Ronald Reagan, Wal-Mart, Amazon and the 1 percent's sick triumph over us all. Salon. http://www.salon.com/2014/06/29/free_markets_killed_capitalism_ayn_rand_ronald_reagan_wal_mart_amazon_and_the_1_percents_sick_triumph_over_us_all/. Last accessed 11th November 2016.

28. Farivar, C. (2014). Uber settles lawsuit with family who sued after 6-year-old killed by driver. Ars Technica. http://arstechnica.com/tech-policy/2015/07/uber-settles-lawsuit-with-family-who-sued-after-6-year-old-killed-by-driver/. Last accessed 11th November 2016.

29. Halavais, A. (2008). Search Engine Society. Cambridge: Polity.

30. Internet Live Stats (2016). Google Search Statistics. http://www.internetlivestats.com/google-search-statistics/. Last accessed 11th November 2016.

31. Lee, J. (2013). No. 1 Position in Google Gets 33% of Search Traffic [Study]. Search Engine Watch. Available: https://searchenginewatch.com/sew/study/2276184/no-1-position-in-google-gets-33-of-search-traffic-study. Last accessed 6th November 2016.

32. Sullivan, D. (2010). Dear Bing, We Have 10,000 Ranking Signals To Your 1,000. Love, Google. Search Engine Land. Available: http://searchengineland.com/bing-10000-ranking-signals-google-55473. Last accessed 6th November 2016.

33. Lovink, G. (2011). Networks without a cause. Cambridge: Polity.

34. Edwards, J. (2014). Here's The Evidence That Google's Search Results Are Horribly Biased. Business Insider. Available: http://www.businessinsider.com/evidence-that-google-search-results-are-biased-2014-10?IR=T. Last accessed 6th November 2016.

34. O'Reilly, L. (2016). The row between Google and review sites is flaring up again. Business Insider. Available: http://uk.businessinsider.com/yelp-tripadvisor-complain-google-must-try-search-burying-sites-results-2016-8?r=US&IR=T. Last accessed 6th November 2016.

35. Moon, M. (2015). FTC report reveals how Google manipulated its search results. Engadget. Available: https://www.engadget.com/2015/03/20/ftc-report-google-search-bias/. Last accessed 6th November 2016.

36. Lasar, M. (2011). Google v. Belgium "link war" ends after years of conflict. Ars Technica. Available: http://arstechnica.com/tech-policy/2011/07/google-versus-belgium-who-is-winning-nobody/. Last accessed 6th November 2016.

37. Mullin, J. (2015). Spain's "Google tax" has been a disaster for publishers, new study shows. Ars Technica. Available: http://arstechnica.co.uk/tech-policy/2015/07/new-study-shows-spains-google-tax-has-been-a-disaster-for-publishers/. Last accessed 6th November 2016.

38. O'Brien, C. (2015). Google's slippery slope: If search giant pays Twitter for content, should it pay all publishers?. VentureBeat. Available: http://venturebeat.com/2015/02/05/googles-slippery-slope-if-search-giant-pays-twitter-for-content-should-it-pay-all-publishers/. Last accessed 6th November 2016.

39. DuckDuckGo (2016). About: Take back your privacy. Available: https://duckduckgo.com/about. Last accessed 6th November 2016.

40. Net Market Share (2016). Desktop browser versions market share. Available: https://www.netmarketshare.com/browser-market-share.aspx?qprid=2&qpcustomd=0. Last accessed 6th November 2016.

The internet improves our ability to communicate and socialise

1. McLuhan, M. (1964). Understanding Media: The Extension of Man. Reading: Cox & Wyman Ltd.

2. Tessandier, A. (2014). Citizens of the Internet. The Huffington Post. Available: http://www.huffingtonpost.com/axelle-tessandier/citizens-of-the-internet_b_4495550.html. Last accessed 6th November 2016.

3. Wurman, R. (1989). Information Anxiety. London: Pan Books Ltd.

4. Postman, N (1985). Amusing ourselves to death. York: Methuen & Co.

5. Stone, B. (2014). Facebook Turns 10: The Mark Zuckerberg Interview. Bloomberg. Available: http://www.bloomberg.com/news/articles/2014-01-30/facebook-turns-10-the-mark-zuckerberg-interview. Last accessed 6th November 2016.

6. Keen, A. (2014). The internet is not the answer. London: Atlantic Books.

7. Zafra, A. (2008). A Friendly Reminder: Facebook is a Social

Utility not a Social Networking Site. AdWeek. Available: http://www.adweek.com/socialtimes/a-friendly-reminder-facebook-is-a-social-utility-not-a-social-networking-site/8122. Last accessed 6th November 2016.

8. Turkle, S. (2013). Alone Together. New York: Basic Books.

9. Pariser, E. (2012). The Filter Bubble: What The Internet Is Hiding From You. London: Penguin.

10. Shapira, I. (2010). Texting generation doesn't share boomers' taste for talk. Washington Post. Available: http://www.washingtonpost.com/wp-dyn/content/article/2010/08/07/AR2010080702848.html. Last accessed 6th November 2016.

11. Axon, S. (2010). 5 Ways Facebook Changed Dating (For the Worse). Mashable. Available: http://mashable.com/2010/04/10/facebook-dating/. Last accessed 6th November 2016.

12. Cialdini, R. (2007). Influence: The Psychology of Persuasion. New York: HarperBusiness.

13. Harris, M. (2014). The End of Absence: Reclaiming What We've Lost in a World of Constant Connection. London: Portfolio Penguin.

14. Quast, L. (2013). Personal Branding 101. Forbes. Available: http://www.forbes.com/sites/lisaquast/2013/04/22/personal-branding-101/. Last accessed 6th November 2016.

15. Lovink, G. (2011). Networks without a cause. Cambridge:

Polity.

16. Mayfield, A. (2010). Me and my web shadow. London: A & C Black Publishers Ltd.

17. Poladian, C. (2014). NameTag: Facial Recognition App Checks If Your Date Is A Sex Offender But Should You Use It?. International Business Times. Available: http://www.ibtimes.com/nametag-facial-recognition-app-checks-if-your-date-sex-offender-should-you-use-it-1539308. Last accessed 6th November 2016.

18. Kickstarter (2016). Here Active Listening - Transform The Way You Hear The World. Available: https://www.kickstarter.com/projects/dopplerlabs/here-active-listening-change-the-way-you-hear-the. Last accessed 6th November 2016.

19. Hooton, C. (2015). This mute button for real life could have some very dark consequences. Independent. Available: http://www.independent.co.uk/life-style/gadgets-and-tech/news/this-mute-button-for-real-life-could-have-some-very-dark-consequences-10316426.html. Last accessed 6th November 2016.

20. Internet Live Stats (2016). In 1 second, each and every second there are. Available: http://www.internetlivestats.com/one-second/. Last accessed 6th November 2016.

A realist conclusion
1. Postman, N. (1992). Technopoly: The Surrender of Culture to Technology. New York: Vintage Books.

2. Titcomb, J. (2016). Internet piracy falls to record lows amid rise of Spotify and Netflix. The Telegraph. Available: http://www.telegraph.co.uk/technology/2016/07/04/internet-piracy-falls-to-record-lows-amid-rise-of-spotify-and-ne/. Last accessed 6th November 2016.

www.ingramcontent.com/pod-product-compliance
Lightning Source LLC
Chambersburg PA
CBHW021405170526
45164CB00002B/513